HOME INVENTIONS

USBORNE PUBLISHING

Acknowledgements
We wish to thank the following
individuals and organizations for
their assistance and for making
available material in their
collections.

Key to picture positions:
(T) top, (C) centre, (B) bottom,
(L) left, (R) right.

British Museum, 22–3(C)
The Design Council, 16(BR)
David & Charles Ltd. from *Yesterday's
Shopping,* 18(CL), 19(CL, CR)
GEC-Marconi Electronics Ltd., 42(T)
Gillette Industries Ltd., 32(BL, T)
Hoover Ltd., 28(TR)
Mary Evans Picture Library, 13(BL),
15(TL), 19(BL), 26(C), 38–9(T), 40(T),
41(TL, CL), 45(C)
Newton, Chambers & Co. Ltd., 15(TR)
Popperfoto, 41(BL)
The Post Office, 38(BC, *second from
left,* BR)
Sperry Remington, 36(TL)
Radio Times Hulton Picture Library,
25(BR)
The Science Museum, London, 10(C),
18(T), 22(BL), 26(BC), 27(BL),
29(TC, BR), 36(B), 37(BL, BC, BR),
38(BL, BC *second from right*),
44(TL. CL); Crown Copyright,
28(TL), 29(BL)
Singer Sewing Machine Company
Ltd., 34(TL)
T.I. Domestic Appliances Ltd., photo
Science Museum, London, 27(BR)
Victoria and Albert Museum; Crown
copyright, 22(T), 23(T)

We would also like to thank the
following groups for their help in the
preparation of this book:

The Science Museum, London
The Design Council
Independent Broadcasting Authority
GEC-Marconi Electronics Ltd.
The Institution of Heating and
Ventilating Engineers
The Deutsches Museum, Munich

Illustrators and Photographers
Roland Berry
Peter Dennis
Terry Hadler
Peter Henville
John Hutchinson
Linden Artists
Peter Mackertich
Mike Roffe
David Slinn
John Thompson
Jenny Thorne

Editor
Sue Jacquemier

Art Editor
David Jefferis

Picture Manager
Millicent Trowbridge

Picture Researcher
Caroline Lucas

Typesetting
Garland Graphics, Bristol

Colour reproduction
Fotolitho Drommel, Zandvoort.
Holland.

Made and printed in England by
Tinling (1973) Limited
Prescot, Merseyside

First published in 1975 by
Usborne Publishing Ltd
20 Garrick Street
London WC2

Text © 1975 by Molly Harrison
Artwork © 1975 by Usborne
Publishing Ltd

ISBN 0 86020 017 5

HOME INVENTIONS

MOLLY HARRISON

Automatic
tea-maker (1902),
with alarm
clock

CONTENTS

4	Beginnings
6	Lighting: from Candles to Electricity
10	Keeping Warm: from Coal Fires to Central Heating
14	Cooking Ranges and Ovens
18	Household Gadgets
20	Refrigerators
22	Sanitation: Drainage and Water Closets
25	Baths, Taps, Geysers and Boilers
28	Vacuum Cleaners
30	Washing Machines
32	Zips, Locks, Razors, Thermos Flasks
34	Sewing Machines
36	Typewriters
38	The Telephone
40	The Gramophone
42	Radio
44	Television
46	"Firsts" of Home Inventions
47	Index

Beginnings

▲ **Roman hypocaust,** or central heating system. From a charcoal furnace, hot air circulated under the floor and up the walls.

► **A central hearth** was the only heating of a medieval home. There was no outlet in the roof above for the smoke.

Hot air passes under the floor, up air ducts in the walls, and out through holes

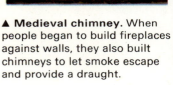

▲ **Medieval chimney.** When people began to build fireplaces against walls, they also built chimneys to let smoke escape and provide a draught.

In pre-historic times, human beings ate only what they could find; life was dangerous and "home" was a dark, cold and often dirty place. In time, families settled down and gradually changed the homes they built and the ways in which they warmed, fed, cleaned and entertained themselves.

Ancient times

Throughout the lifetime of mankind there have been large and powerful civilizations which have left us very little evidence of how they lived. Other ancient peoples have left us more evidence, like the Egyptians who invented such mechanisms as locks and water clocks.

We know quite a lot about the life of a rich Roman family from the ruins of Pompeii. This flourishing city was suddenly and completely destroyed in A.D. 79 during an eruption of Vesuvius, but many of the buildings and possessions of the inhabitants were preserved in the rubble under the hot lava. Prosperous homes had mosaic floors and underfloor heating from a hypocaust, a kind of

furnace from which hot gases were circulated in pipes through the house.

The population of Rome itself grew fast and over-crowding was a great problem. Entire families were often herded together in one room, amid rubbish and filth. They did not have the water supplies and drainage enjoyed by the richer people. Accidental fires were frequent because of the movable stoves, candles, torches and smoky lamps which were used by everyone.

The Middle Ages

In the Middle Ages life was violent and dangerous. Throughout Europe there was constant fighting between rival barons, living in castles which were cold, dark and draughty. There were no separate rooms and everybody lived, ate and slept in the same large room, called "hall" in English, *salle* in French and *saal* in German.

The doors and "wind-eyes" (windows) were small and open to the weather. The floor was of trodden earth

▲ **Open hearths** were in use for centuries. Pots hung from a chimney "crane" and meat roasted on a spit.

▼ **Tinder box.** The flint was struck against the iron to make a spark, setting fire to the cloth in the tin; spills were then lit from the flame.

Flint

Iron

Charred rag

Candle was lit with spills

Spills

▲ **Pumps and wells** were the only sources of water before proper drainage which began in the 19th century.

Rush-light could be lit at both ends

▲ **18th century rush-lights,** made from stalks of rushes, gave a faint, flickering light. This iron "nip" incorporates a candle-holder.

▼ **18th century horn "books" and alphabets** were used by children. The paper was covered with a sheet of horn, nailed or sewn down.

and the only heating was by a bonfire of wood, in the middle of the room, and everyone slept on the floor.

The 16th and 17th centuries
Gradually people began to be more concerned about privacy and comfort in the home. By the 16th century thick walls and tiny windows were old-fashioned and the homes of the wealthy were friendly places, with many rooms and large windows, wall fireplaces and tall chimneys. Even cottages were being improved: windows were no longer filled in with pieces of horn or leather, but with small panes of glass joined with strips of lead.

The 18th century and after
Until the 18th century, nearly all work was either farm work or "cottage industry" – men, women and children making goods in their own homes. Many of the cottages were dark, damp and crowded, but the people in them were part of a strong and stable village community.

A change in this way of life first came about in

England. Landlords began to interest themselves in improving farming; scientists were discovering new forms of power; and there was a great surge of inventive genius which affected domestic appliances and home life in general. By the end of the 19th century, railways were transporting goods more easily than ever before. This was the time of the first Industrial Revolution, when people no longer made things by hand, but by machines in factories. It gradually altered everybody's work and leisure and the kind of homes they lived in.

This book is about inventions which have improved everyday life since 1750, and about ideas for changing the home which grew out of the first Industrial Revolution. It is also about domestic inventions of the last 75 years, in which the world has gone through a second industrial revolution, bringing us electricity, the motor car, the aeroplane, radio and television. We are now on the threshold of yet a third revolution – that of electronics, atomic power and super-conductors.

Lighting: from Candles to Electricity

Before gas or electric lighting, most people got up as soon as it was light and went to bed when it got dark. They had to manage with the flickering light of candles or rushlights. Tallow candles, made of mutton or pig fat, were smelly and melted very quickly and unevenly. As living standards improved in the 18th century people began to want better lighting. Beeswax was used for candles and gave a cleaner and pleasanter light. Wicks which were plaited instead of twisted made candles burn with less mess, so they no longer needed to be trimmed or "snuffed" regularly. In large houses there was a special "lamp room" where the lamps were kept and cleaned and where moulds for making the candles were stored.

Oil lamps had been used for centuries, but not until the invention of the Argand burner was any successful attempt made to improve the size of the flame.

▼ **Roman oil lamp** of clay. A wick made of linen was laid in the channel with one end in the oil. The other end was then lit.

Reservoir for oil

Green shade to reduce glare

Glass chimney

Screw to adjust height of lamp

Wick adjuster

▲ **A reading lamp with an Argand burner.** The Argand lamp was invented by Argand, a Swiss, in 1784. It was the first big change in lighting since early times. It had a new kind of burner which allowed more oxygen to be drawn up inside the wick, increasing the current of air and giving a better light. There was a wick adjuster, a glass chimney to avoid draught and give a steady light.

▶ **The Argand lamp** on the right had a reservoir for the oil which was fed to the burner around the wick by gravity. The level of oil was kept constant by means of a valve at the lower end of the reservoir. These lamps used five times as much oil as the simple oil lamps so they were expensive to run.

GETTING MORE LIGHT FROM CANDLES

▼ Four ways of grouping candles. Wall sconces (**1**) were fittings with sockets for candles and pans to catch the grease. Chandeliers (**2**) hung from the ceiling and held either a few or dozens of candles. A girandole (**3**) was a wall mirror with candles round it so that the light was reflected. Sometimes cabinets or writing desks were fitted with special pull-out shelves (**4**) to stand a candle on when extra light was needed.

1

2

3

4

Chimney stops draughts from blowing out flame

Decorative glass globe

"Galley" supports globe and chimney

Wheels raise and lower wicks

Paraffin reservoir, made from glass

Heavy base: a typical late 19th-century design

A new kind of lighting in towns

Gas lighting was first used in factories as early as 1792 and Pall Mall, in London, was the first street to be lit by gas, in 1807. It was not until about 1840 that gas lighting in town houses became fairly general. The gas pipes in each room led to a central plaster "rose" in the ceiling and to S-shaped brackets on either side of the fireplace. The flame was enclosed in a frosted glass globe and gently popped and purred as the gas escaped. The light was provided by a jet until the invention of the incandescent mantle *(see page 9)*, which was not really safe to use until the 1880s.

Paraffin lamps

Paraffin — at first called kerosene — was found in great quantities in Pennsylvania in 1859 and was soon being exported to Europe. This mineral oil was cheaper than vegetable or whale oils had been and caused a revolution in home lighting. The paraffin lamp was simple to use, reasonably safe, and gave a steady light almost without smell. Now, for the first time ever, country homes could be as well lit as those in towns.

◄ **Paraffin lamp** with a "duplex" (double) burner, late 19th century. The two parallel flat wicks were easy to trim and were raised or lowered into the oil by two wick wheels. The tall base put the light at a good level for reading.

▼ **An experiment in making gas** from coal was carried out in the early 18th century by an Irishman, the Rev. Clayton. He filled a kettle with coal, heated it and lit the gas which came out of the spout.

Extended spout

Coal, when heated, gave off gas.

▼ **Different gas burners** gave different flames. **1.** Brass burner of 1812. **2.** "Batswing" burner 1820–30. **3.** Double-flamed burner of 1825. **4.** Six-flamed burner of 1858; the flames came through soapstone holes which did not corrode.

1

2

3

4

Gas lighting rivalled by electricity

Gas lighting was one of the first inventions which had the effect of reducing domestic work. People no longer had to snuff candles, trim the wicks of oil lamps, or keep their lamp reservoirs filled regularly with oil. If you could afford to have gas in your home all you had to do was to light it. However, there were still drawbacks: the open flame of a burner deposited carbon particles on ceilings, walls and curtains.

Gas lighting would probably have developed further, but towards the end of the 19th century a cleaner and even brighter kind of lighting was developed: electricity. Many people had produced electric lamps from 1845 onwards, but none of them was practical until the English scientist Swan, in 1878, and the American Edison, in 1879, both produced lamps which would last for as long as 150 hours. A quarrel developed over who was the first with the invention, but fortunately the two men settled their differences and together they set up the Edison-Swan Electric Company to manufacture light-bulbs. It is still in existence today.

At first, if people wanted to have electricity they had to install their own generators, but gradually public supplies were provided. Electric lighting gave off much less heat than gas had done and it was safer. For a time, however, it was very expensive and the "battle" between gas and electricity lasted for many years.

Neon lighting

"Neon" lights, known to scientists as electric discharge lamps, were first made around 1910. These lamps are glass tubes which have no filament. The air is pumped out and the tube is filled with gas: mercury, helium or neon gases can be used. When electricity is passed through the tube, this causes the gas to give off a particular colour, and a bright light.

Fluorescent lights *(see opposite)* have a mercury gas inside the tube, but also a fluorescent powder, which glows brightly when electricity is passed through the tube. Each powder has its own colour, and by a careful selection of different powders, it is now possible to make fluorescent lights which give off almost any colour.

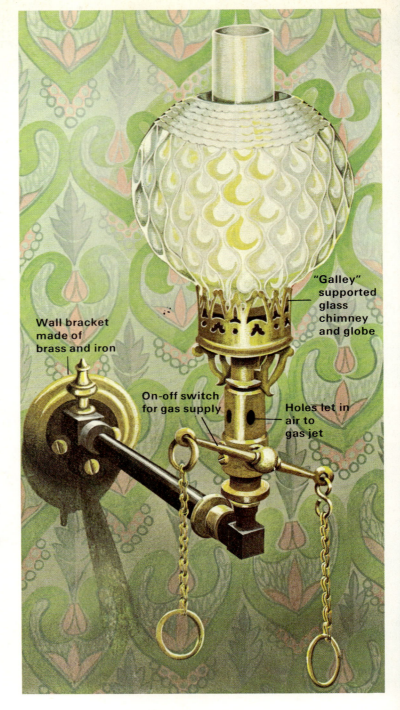

"Galley" supported glass chimney and globe

Wall bracket made of brass and iron

On-off switch for gas supply

Holes let in air to gas jet

THE DEVELOPMENT OF THE MATCH
Getting a light was a slow, difficult and often dangerous business, but a number of chemists made improvements in the 19th century.

1. In 1826 John Walker made "friction lights" – thin, flat sticks of wood with a paste on one end

Pipeclay rod ────

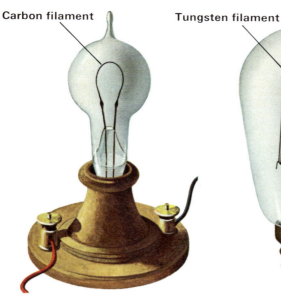
Carbon filament

Tungsten filament

Coiled coil

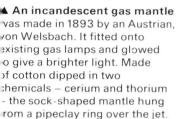

▲ **An incandescent gas mantle** was made in 1893 by an Austrian, von Welsbach. It fitted onto existing gas lamps and glowed to give a brighter light. Made of cotton dipped in two chemicals – cerium and thorium - the sock-shaped mantle hung from a pipeclay ring over the jet.

◄ **Wall gas lamp** of the type which could be fitted with a mantle.

► **The modern fluorescent "tube" fitting** is cheap to run and casts a bright, even light. The tube is filled with low-pressure mercury gas. When the electricity supply is turned on, electrons travel down the tube (from **A** to **B**), completing the electrical circuit. Electrons collide with mercury atoms, which give off ultra-violet light (invisible to our eyes). This makes the coating glow, or fluoresce".

▲ **The Edison electric lamp** of 1879 was a glass globe containing a carbon filament in a vacuum. He made the filament by baking a piece of ordinary sewing cotton. The bulb would not have been possible without the invention in 1865 of a vacuum pump, nor without the introduction of new ways of glass-blowing and of fusing metals.

▲ **The Tungsten bulb.** Other kinds of electric filament lamps followed those using carbon. Tungsten could be run at higher temperatures than other metals but was at first too fragile. Coolidge in the United States found how to make tungsten tough and the tungsten vacuum lamp of 1911 was stronger, less expensive, and gave out a better light than the carbon filament.

▲ **A modern electric bulb** has a filament made of tungsten, coiled to give more light. The total length of the filament in a 15 watt bulb is about 27 ins. (0.75 m.). The first coil has about 3,000 turns and is recoiled in 100 larger turns. Since about 1925 electric bulbs have been frosted on the inside to reduce glare.

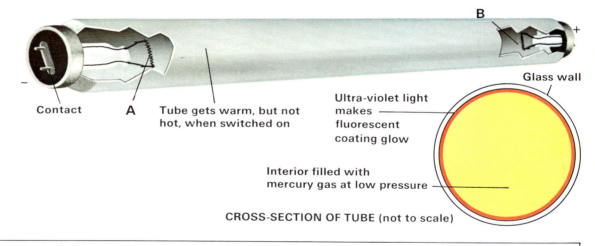
B
+
Glass wall
Ultra-violet light makes fluorescent coating glow
Contact   A   Tube gets warm, but not hot, when switched on
Interior filled with mercury gas at low pressure

CROSS-SECTION OF TUBE (not to scale)

which burst into flame when rubbed against sandpaper. **2.** "Lucifers" had sulphur heads and **3.** Punch's "Congreves" had heads of sulphur, and sandpaper on the bottom of the box. **4.** "Vesuvians" were double-ended matches made in 1880.

5. "Vestas", made about 1873 in Italy for sale in China, were made of cotton dipped in candlewax. **6.** "The Rupee", a Norwegian match, had heads of yellow phosphorous. Safety matches were made in Sweden in 1855.

4
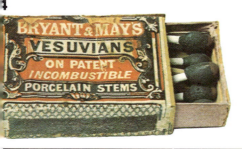

5

6

Keeping Warm: from Coal Fires to Central Heating

Before the late 19th century, the way to heat a single room was to light an open fire or a closed stove. The open fires had grates, made for burning coal, which stood on legs to encourage draught. Because hearths were usually very wide and chimneys huge, most of the heat was wasted.

A British inventor, Count Rumford, who was also a social reformer, wrote a paper in the 1770s on smoky chimneys. In it, he put forward some ideas for increasing the efficiency of fireplaces, using special insulated firebricks and a narrower hearth. As a result, the traditionally large, open fireplaces in many houses were "Rumfordized", to the delight of brickmakers and to the dismay of iron grate-makers.

Various kinds of large metal or tiled wood-burning stoves were used widely in the colder parts of Europe and America, as they had been for centuries, and in many places still are. The room in which the stove was kept was itself often called a stove or "stew" (*étuve* in French, and *stufa* in German).

The idea of heating a room by circulating warm air came originally from the Romans, but it was applied to 18th century houses by an American statesman, Benjamin Franklin, in 1740. The fire itself was covered, and air entering the stove at the front was warmed, emerging at the back of the stove through air vents *(see page 12)*. Wood was used for fuel in these stoves, since forests were plentiful in the new lands.

Gas fires and paraffin heaters

No alternative to coal or wood fires and stoves was developed until the 1870s when the first gas fires appeared. These were mainly due to the work of Baron Bunsen in the 1850s; he developed a new type of gas burner. They worked by heating with gas flames a piece of fireclay or asbestos until it was glowing, so that heat was radiated out into the room. Early gas fires had no flues and often exploded.

Paraffin heaters, sometimes called kerosene lamps, were developed from paraffin lamps and have continued to be a popular kind of room heater.

Electric heaters

The earliest electric fires consisted of flat plates of cast iron heated internally by wires carrying an electric current. Various kinds of wire were tried out and nickel-chrome wires began to be used in 1906. They made the fires more reliable, because they did not go rusty even when red hot. Electric storage heaters use cheap off-peak current to heat bricks or concrete blocks which are inside a metal casing. These heaters retain heat for a long time and have been widely used in Europe since the 1950s.

◀ **Portable gas fire** of 1890 by Welsbach (who also invented the gas mantle for lamps). He had found that, when heated, fireclay glowed with heat more effectively than asbestos and his fire had five perforated radiants made of fireclay standing over gas jets. The case was of enamelled cast iron and a water vessel in front of the radiants helped prevent the air in the room from getting too dry.

▶ **Gas fire** of 1882. It was not until the 1880s that tufted asbestos was found to be suitable as a radiant and gas fires of this type began to be sold in large numbers.

▼ **Open grate,** about 1770. The huge chimney-pieces were very wasteful of heat. On the right are some improvements.

▼ **Rumford** showed how to save heat. He reduced the size of the fireplace and chimney opening and sloped the walls outwards.

▼ **Portable oil stove,** 1890, in ornamental cast iron, with Gothic-style glass windows. The top could be removed, to reveal a cooking plate.

▲ **Belling's electric fire** of 1913 had a place to hang a kettle and a rack for toasting.

▼ **How modern electric fan-heaters work.** Cold air is drawn in by rotary fan-blades which are placed over a series of heating coils. The air is warmed up and is then blown out into the room.

▼ **German earthenware stove.** In cold countries closed stoves developed from open fires; they were hotter and smokeless.

▼ **Hall stove** of cast iron, with a flue pipe carrying the gases out of the room. The surface of the flue gave off heat.

Heating the whole house

The idea of making one fire heat several rooms had been used successfully by the Romans, but was then forgotten. In the 18th century inventors began to work out ways of circulating either hot water or steam throughout a whole house. By the middle of the 19th century large iron boilers, burning wood or coal were being installed in a convenient central place in large houses, and iron pipes were run along the walls to send the water or the steam to large radiators. These were shaped so as to give out as much heat as possible and were often elaborately decorated.

Some people objected to the smell the early heating systems caused; others thought the iron pipes spoiled the look of the rooms. Steam systems were disliked particularly because the temperature was difficult to regulate and condensation of steam in the pipes caused water-locks and made a lot of noise, called "bumping" or "hammering".

Gas, oil and electricity

As gas installations gradually became more efficient they were used for central heating. Oil was thought to be dangerous for home use at first, but is now common.

A newer way of using electricity is by under-floor heating, which can be installed very cheaply in a new house, but not put into one already built. Heating cables are embedded in the concrete of the ground floor and the system can be time-switched to use only off-peak current. Electric heating can also be embedded in plastered walls and very small elements can be put into a new type of plastic wallpaper.

District heating

The idea of heating several buildings from one central plant originated in the United States in the 1870s. This is called "district heating" and many European cities are developing the principle, using spare steam from power stations or gas given off by burning rubbish.

Conserving heat

Nowadays methods of saving heat are important for two reasons: firstly to save money and secondly to conserve fuel, which will become more scarce as the world population increases and resources dry up. To avoid loss of heat in a house, the roof, ceilings and walls can be lagged, draughts can be excluded and windows can be double-glazed.

Air conditioning

As early as 1911 Willis Carrier, an American, was studying the principles on which modern air conditioning units are based. The idea of "conditioning" air to meet special needs, such as in factories, theatres and restaurants was gradually extended to apply to the home. Air can now not only be heated or cooled, but also cleaned till it is free from dust, pollen, soot, smells and other impurities. Winter systems clean, heat, humidify and circulate the air, while summer systems cool and dehumidify, clean and circulate it.

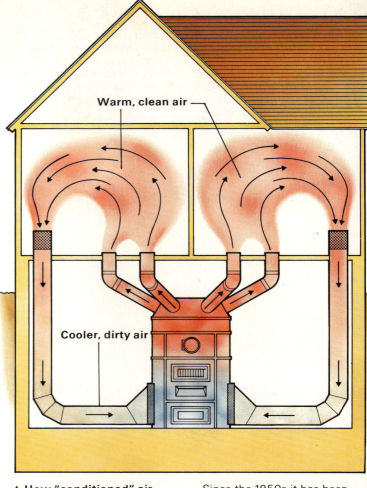

▲ **How "conditioned" air circulated from a furnace in the basement of a house, about 1947.** Air conditioning began in the United States early in this century, as a way of regulating the temperature and moisture of the air in factories.

Since the 1950s it has been developed so that a circulation system will filter air coming into a house, heat or cool it, make it dryer or moister and expel it.

SOURCES OF HEAT FOR CENTRAL HEATING SYSTEMS

▼ **The Franklin stove** of 1742. A flue at the back formed a kind of radiator. Warm air passed over it, and into the room. The smoke went up the chimney.

▼ **Central heating** in 1745. An early idea by Colonel Cook. He thought of installing an engine to heat eight rooms of a house by steam pipes fixed to the walls like radiators.

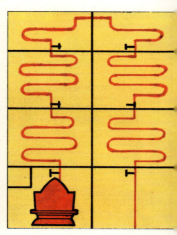

▼ How hot water central heating works. This diagram shows the basic principles of the small-bore central heating system, using pipes of ½ in. (12mm.) diameter or less. The water heated by the boiler is circulated around the house by an electric pump, through the narrow pipes. The water in the radiators gives off heat to the rooms, and then returns to the boiler for re-heating. The bright red areas in the diagram indicate hot water; the brown areas indicate cooler water. The arrows show the direction in which the water flows inside the pipes and radiators.

In earlier designs the water circulated through the system by convection. Hot water rose upwards through the system, and cold water sank to the boiler which had to be at a low level – often in the cellar. Nowadays an electric pump controls the water, so the boiler need not be below ground level.

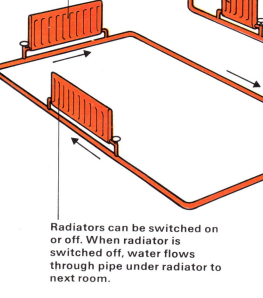

Cold water storage tanks

Extra cold water is fed into system when water is occasionally bled off from radiators

Heated towel rail

Hot water storage tank

Radiator

Electric pump

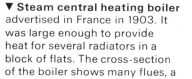

Radiators can be switched on or off. When radiator is switched off, water flows through pipe under radiator to next room.

Cooler water flowing back to boiler to be reheated and recirculated

Thermostatically-controlled boiler

▼ Steam central heating boiler advertised in France in 1903. It was large enough to provide heat for several radiators in a block of flats. The cross-section of the boiler shows many flues, a large furnace and wide pipes, leading off the boiler to the floors above. Notice the decorative design of the radiator.

▼ "Ideal" domestic boiler, 1930, burnt coke or anthracite, and claimed to be "neat in appearance and suitable for fixing in a hall or living room".

▼ Modern solid fuel boiler, from which a central heating system can be run. Like many modern appliances, this boiler is enclosed within a casing, which hides the furnace.

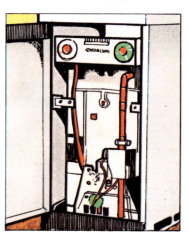

Cooking Ranges and Ovens

In the 18th century, with the more widespread use of coal for heating, narrow grates began to replace the wide, open hearth. It was then no longer possible to cook large joints of meat on horizontal spits, and an improved kind of cooking mechanism, called a "jack" was invented, which was turned by clockwork. The joint was hung up in front of the fire by a chain connected to a brass cylinder which enclosed a spring. The key for winding this up hung nearby and, when the mechanism was started, the joint revolved as it dangled from the "jack".

Cooking utensils made of cast iron had always stood on the floor by open fires and were heavy and awkward to lift. Gradually it was realized that it would be much easier if they could be kept on the same level as the fire. A grate developed which had a flat surface, called a "hob", built on either side, on which pots could stand.

Cooking ranges

There were criticisms about the loss of heat and waste of fuel in open hob grates and experiments were made with a hot-plate – a long, flat, cast-iron plate set on a brick base and with a fire box under it. In 1780 a patent was taken out by Thomas Robinson for the first kitchen range. (A patent is a licence granted by a Patent Office to an inventor, to prevent, by law, anybody else from making and selling the invention which he has "patented"). The range had a cast-iron oven on one side and a boiler for heating water on the other. The oven was lined with bricks and mortar. It was warmed by contact with the fire on one side. Later, an iron rod was used as a conductor between the fire and the oven.

The fire was still open in this new kind of cooker and it was not until 1802 that a closed-top cooking range was invented. The front of the fire was still open for toasting and sometimes the oven was raised and set at the back of the hot plate, but this meant that the cook had to reach across the fire. In America, the closed ranges were small enough to be moved from place to place. They first appeared in the 1830s; prior to this the early settlers used open fires and brick ovens for cooking.

Closed ranges were soon very popular and gradually came to be used in most houses. They were very extravagant with coal, but this was still a cheap fuel because miners were paid so little. Most ranges made a great deal of smoke and soot and had to be blackleaded and brushed every day. Many of these "kitcheners", as the closed ranges came to be called, burnt very fiercely and were very difficult to control. However little cooking was being done, the fire could not be reduced much unless the cook was very skilful in removing some of the front bars and part of the hot-plate. Many housewives began to compare the dirt and the expense of the wasteful coal ranges with the costly but very convenient new gas cookers.

▲ **An open-fire range of about 1850,** with an oven at one side and a back boiler which had to be filled and emptied by hand. Later, boilers were put at one side of the fire and the water could be drawn off by a tap.

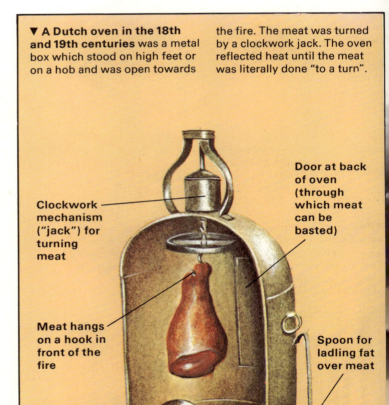

▼ **A Dutch oven in the 18th and 19th centuries** was a metal box which stood on high feet or on a hob and was open towards the fire. The meat was turned by a clockwork jack. The oven reflected heat until the meat was literally done "to a turn".

Clockwork mechanism ("jack") for turning meat

Door at back of oven (through which meat can be basted)

Meat hangs on a hook in front of the fire

Spoon for ladling fat over meat

Tray to catch dripping fat

▲ **An American range of 1888,** with a large hotplate, an oven to one side and a water boiler, with a tap, on the other. The heat was regulated by closing or opening the doors in the front. The flue was fitted into the old chimney.

▼ **The Aga cooker was invented in 1924 by Dr Gustav Dalen,** the Swedish physicist and Nobel Prize winner. The solid fuel burns at about 900°F (480°C) in an enclosed cylindrical furnace, surrounded by heavy metal. All parts are heavily insulated; a thermostat keeps the temperature constant, by controlling the amount of air which enters the furnace. The heavy lids keep the boiling plates always hot. Some models have a water heater alongside; newer ones burn gas or oil.

▲ **A kitchener of 1875,** with four separate roasting and baking ovens, grillers, hot "closets" for dishes and a boiler with tap. It was made of iron with steel and brass fittings.

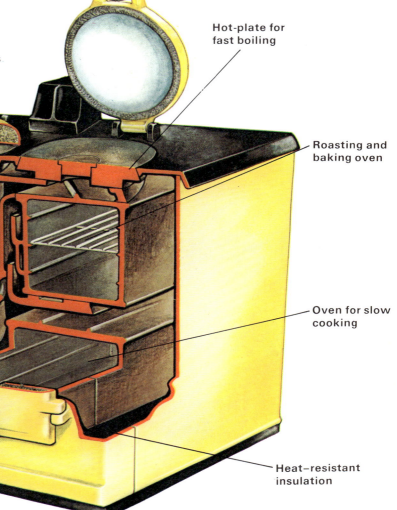

Thermostat (controls temperature)

Enclosed fire unit

Water heater

Fire-grate (rotates fuel)

Hot-plate for fast boiling

Roasting and baking oven

Oven for slow cooking

Heat–resistant insulation

Cooking by gas

Experiments in cooking by gas were begun in the 1830s but it was thought that impurities in the gas would spoil the taste of the food. By the 1850s gas cookers, or gas "cooking ranges" as they were called at first, were being sold. They were smelly, noisy and dirty and very few people bought them. Compared with the growth in popularity of gas for lighting purposes, the adoption of gas for cooking was surprisingly slow. Gas companies began to hire out their cookers in the 1870s and they gradually became more popular. Around 1880, a manufacturer in Chicago was happy to report that "the popular prejudice [against gas] is gradually giving way".

Cooking by electricity

The first electric cookers were shown at an Electric Fair in London in 1891. They had very unreliable heating elements and were slow and expensive to use. Until about 1912 most electric cookers were merely converted gas cookers. Many people were terrified of being electrocuted by the strange new equipment. When servants were plentiful, before World War One, little attention was paid to kitchen apparatus. But when servants found better-paid jobs, the ladies who now had to do their own cooking began to think of ways of saving time and labour. Electric cookers began to achieve widespread popularity in the 1930s as electricity became generally more available.

▲ **Wood-panelled gas oven, made in 1850.** It was still being used by relations of the inventor in 1949. The frame had a cast-iron lining and the casing was packed with fire-clay. The gas burner was on the inside of the oven door.

▼ **Alfred King's gas cooker** was specially installed in the Rothschild house in London in 1859. It had no oven, but a hotplate with three burners, worked by three brass taps. The shape was very similar to that of the old open ranges.

▶ **Mid-19th century gas cooker of cast iron,** with four hotplates and a griller. The oven was lined with white enamel, had a hook from which joints could be hung, and a glass-panelled door. Such cookers stood free, so the control taps were at the side.

▲ **Early electric stoves,** for sale from 1894, had an oven and a warming plate. The oven was heated by three electric elements, controlled by three large switches.

▼ **Mobile electric capsule kitchen** of 1968 with oven, kettle, cooking rings, refrigerator and storage space for crockery and pans. It only needs to be plugged in.

New inventions

A regulating device for setting the oven heat at a required temperature, introduced in 1923, was a great improvement to gas stoves. In 1931, an electric oven thermostat was also introduced.

The microwave oven, patented in England in 1959, is an entirely new way of cooking. It uses high frequency electro-magnetic waves which penetrate directly to the centre of the food. A time-switch controls the cooking time, which is remarkably short: a sausage can be cooked in 10 seconds, a cutlet in 5 and a chicken in 48 seconds! As a time-saver, it is ideal for women who work outside their home as well as in it.

Size, as well as speed, has become an important part of the design of stoves. As houses and apartments in cities become smaller, the need to economize on kitchen space has also become more apparent. The capsule kitchen on the right seems to be the answer to this problem. It can be wheeled from room to room, and, if it has a long enough flex to connect it to the electricity supply, even into the garden.

Household Gadgets

In a modern house the kitchen is very carefully planned and is often thought of as the most important room of all. Since very few people can now afford to keep servants, new kitchens are small and labour-saving, with gadgets of all kinds available to make housework easier.

Comfort and convenience in the kitchen were not much considered, even a century ago. All the pans, hobs, trivets, toasters and other cooking gadgets used for centuries before the Industrial Revolution seem to us to be heavy and clumsy. Before light-weight metals were introduced, saucepans were usually made of thick iron or copper, often with brass fittings.

The new gadgets, which began to be made in the United States in large numbers towards the end of the 19th century, may seem odd to us, for they were often elaborately decorated and very different from the streamlined objects we use today. Many of them cut the time taken to prepare food: things like meat mincers and vegetable choppers. In America, mechanized apple parers were in great demand in the days when apple trees grew around every farmhouse. All these gadgets were welcomed as valuable labour-saving devices by those housewives who could afford them.

Ironing clothes

Ironing used to be a hot and difficult task, achieved by heating up a lump of metal and putting it inside a hollow "box"-iron. Later, ironing was done by heating two flat-irons on a hob and using them alternately. Gas irons were introduced in the middle of the 19th century and electric irons in the early days of domestic current. They were connected to the gas or electric supply from chandeliers, and for the first time made it possible to iron continuously.

Automatic tea-makers

Tea and coffee machines have been improved over the years, and some of the different models are shown here. The automatic tea-making machine is a time-saving device, and is particularly ingenious in its design. Patented in 1902 by a Birmingham gunsmith, it works by a combination of levers and springs. When the alarm goes, the machine strikes a match, lights a spirit stove and boils the water in the kettle. It then automatically pours the water, switches off the stove and rings a bell to say the tea is made!

Electric gadgets

Electricity has become the regular substitute for a domestic servant. By the end of the 19th century many manufacturers' catalogues listed and illustrated most of the electric gadgets which we know today: kettles, saucepans, frying pans, toasters, hotplates and coffee grinders. One electric gadget which still had not been invented then is the extractor fan, which drives the smell of cooking out of the kitchen.

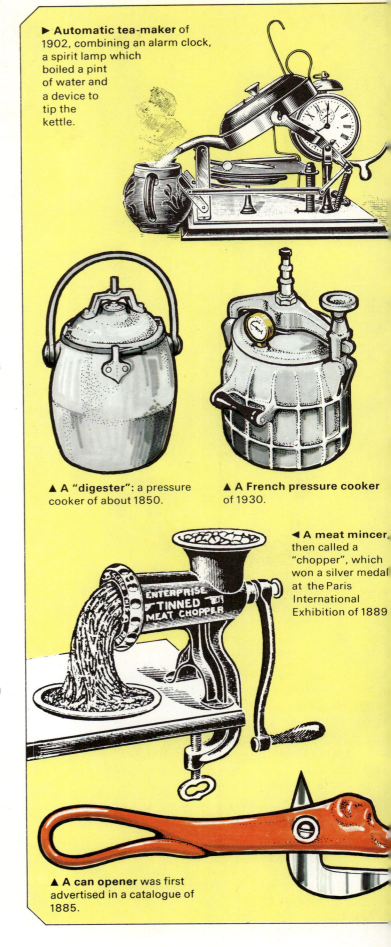

▶ **Automatic tea-maker** of 1902, combining an alarm clock, a spirit lamp which boiled a pint of water and a device to tip the kettle.

▲ **A "digester":** a pressure cooker of about 1850.

▲ **A French pressure cooker** of 1930.

◀ **A meat mincer**, then called a "chopper", which won a silver medal at the Paris International Exhibition of 1889

▲ **A can opener** was first advertised in a catalogue of 1885.

▼ **An egg beater** from a kitchen of 1870.

▼ **An electric mixer** of 1920.

▼ **An electric mixer** of 1966 which beats, blends and mixes food.

▲ **Coffee infuser** of 1890. An early syphon type, it was heated by a spirit stove.

◄ **"Magnet" electric toaster** of 1920. The toast was held in position by two spring clips.

▲ **A copper tea kettle** and stand advertised in 1907, with a paraffin heater underneath.

▼ **A spirit-heated iron,** shown in a store catalogue of 1907.

► **A coffee machine,** from an advertisement of 1907. It was made of polished brass with a glass cover and was heated by a paraffin stove underneath.

▼ **An American gas iron,** 1884. It was like a box-iron, heated inside by a Bunsen gas jet.

► **A modern steam iron,** filled with distilled water and heated by electricity. The garments are damped as they are ironed and the heat can be set to a suitable temperature.

Refrigerators

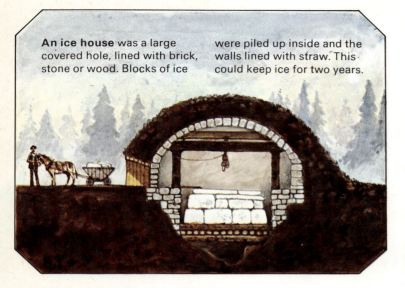

An ice house was a large covered hole, lined with brick, stone or wood. Blocks of ice were piled up inside and the walls lined with straw. This could keep ice for two years.

An ice store, from which fishmongers or street-sellers could buy ice for re-sale to those who had no ice house of their own or whose supply had run out.

An ice box was a lead-lined wooden cabinet, with compartments for storage, drainage holes, and a tap for drawing off surplus water.

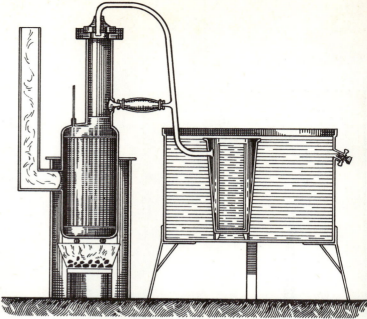

▲ Carré's ice machine of 1860 consisted of a small portable stove, a boiler three-quarters filled with ammonia, a freezing vessel and a reservoir. It needed one hour of heating and one hour of freezing to produce a kilogramme of ice.

Ice-houses

Nowadays, when refrigerators and deep freezers are widely used, it is not easy to realize that before this century the only way in which anybody could have jellies or cold drinks in summer was by storing ice in winter in underground ice-houses. These were filled after a spell of cold weather, when rivers, canals and ponds were frozen over. Only grand houses had these arrangements, but if anybody in the district fell ill, ice was supplied to them if the doctor wrote to say that this would help them to recover.

It had been known for many hundreds of years that food could be kept fresh if it were packed in ice, but the principle of refrigeration was discovered only in the middle of the 19th century. James Harrison, a Scottish printer, developed a machine in Australia after he noticed that ether had a cooling effect when it evaporated. His invention was shown at the International Exhibition in 1862. The idea could also be applied to ships' cargoes and the first cargo of frozen meat was sent to England from Australia in 1880.

Ice-making machines

At first the domestic refrigerator was simply a lidless box containing a lump of ice, but machines which actually *made* ice were developed from the work of a Frenchman, Ferdinand Carré. In 1890 an Englishman, Ash, patented a refrigerator which was a wooden drum with a closed top and a handle which worked two pistons inside a cylinder. Ice and salt were packed round this, with a little water, and the pistons were worked up and down, causing the freezing agents to pass round the cylinders. Later, patent freezing powders were used instead of ice and salt.

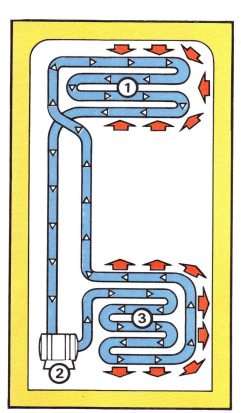

▲ **A refrigerator** works by allowing a liquid cooling agent to evaporate in the pipes inside the fridge. This vapour absorbs heat from its surroundings, thus cooling the fridge (**1**). The vapour is then compressed (**2**), and goes into the condenser (**3**) where it is turned back into a liquid and gives off its heat into the room. The cycle then continues.

Key to diagram

1. Evaporator (in freezer compartment).

2. Automatic switch turns light on when door opens.

3. Star markings. A three-star freezer compartment can store food for up to three months.

4. Flexible plastic door-seal.

5. Magnetic door-catch.

6. Thermostat (regulates temperature).

7. Insulated plastic inner casing.

8. Electric compressor unit.

9. Condenser (outside the fridge at the back).

▼ **A modern refrigerator of the compression type,** cut away to show the pipes which carry the cooling agent (ammonia, ethyl chloride or Freon).

Sanitation: Drainage and Water Closets

The problem of what to do with human sewage was much worse early in the 19th century than ever before. Populations were growing fast and, in England, poor country families were flocking into the towns for work in the new factories.

Houses were urgently needed for them and were often hurriedly built of poor materials and without proper ventilation or drainage. In the bigger towns drains were sometimes built, but they did more harm than good because they emptied into the rivers and streams which provided drinking water.

Even the best houses still had cesspits and these were rarely emptied. When a cesspit was full it was often covered over and forgotten and another one dug nearby. Drinking water from wells was often contaminated by leaky or overflowing cesspits and there was no regular supply of fresh water to houses for cleaning or flushing drains.

Drains and disease

The death rate in towns all over Europe was very high among both rich and poor. Nobody knew anything about germs at this time, or realized that they lived and spread in dirt. The biggest killer of all was cholera and tens of thousands of people died in epidemics of this disease every year. In addition, there were regular outbreaks of typhoid and enteric fever.

Many reformers tried to get the dreadful sanitary conditions improved. In 1848 the first Public Health Act in England gave local authorities powers to get rid of cesspits and to build good sewage systems with the new glazed stoneware pipes, but it was not until the 1870s that the death rate began to fall.

▲ **Part of the trade card** of a "nightman" who emptied cesspits of London houses in the late 18th century.

▼ **A drawing of 1845,** showing sewers being built in London. Glazed stoneware pipes began to replace iron ones.

▼ **Harington's water closet** with a flushing cistern, 1596. It drained into a "vault" which could be emptied at night.

▼ **Cumming's Closet of 1775** had a U-bend with water. This cut off the pan from the cesspit and prevented any smell.

▼ **In 1778 Bramah re-designed the closet.** His model had two hinged valves, one to let water into the bowl, and one to drain it. Thousands were sold well into the 19th century before the introduction of wash-out closets.

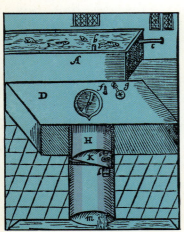

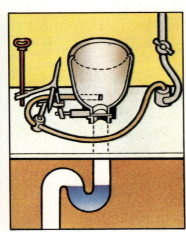

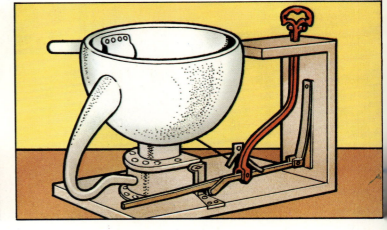

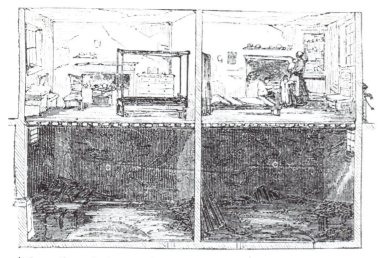

▲ **A section of a house** showing cesspits under the living-room floors. Even in the best houses this arrangement was usual.

The development of the water closet

It is amazing what human beings can put up with if they have to. Until relatively recent times everyone – rich and poor alike – was accustomed to living in dirty conditions and with unpleasant smells which we, nowadays, would find intolerable. In the country human excrement was used for fertilizing the land but in towns it either lay about or was thrown into streets or rivers.

When the first water closet was invented by Sir John Harington at the end of the 16th century it was not taken seriously. It was a crude affair and to make it work would have needed a supply of running water which very few towns had at that time. But it shows us that men's minds were beginning to be concerned with problems of hygiene and health.

The 18th and 19th centuries

No improvement was attempted for the next 200 years and people continued to dispose of their sewage as they thought best, without bothering about their neighbours. In the 18th century various kinds of water closets with fitted valves were made, but for a long time these were very expensive and they were unpleasant to use because they were not ventilated. In 1775, Alexander Cummings, of London, took out the first patent for a water closet.

During the 19th century, many people took out patents for improved kinds of water closets. There were several problems which needed to be solved. A method had to be found of flushing the bowl quickly and thoroughly, without too much noise. The inventors had to make sure that the apparatus was easy to keep clean and not too expensive.

There were several basic types: the pan closet, an early and very unhygienic device; the valve closet (developed on the same principles as Harington's closet); the hopper closet, which was highly inadequate, but considered good enough for servants; the wash-out closet, a popular type made in the late 19th century; and the wash-down closet *(see page 24)*. A further type, the syphonic closet, was popular in the United States.

▼ **Moule's earth closet** of 1860 released earth or ashes from a hopper above the seat, allowing them to fall into a bucket.

▼ **The wash-out closet** of 1870 had an overhead cistern. A small amount of water was left in the pan to act as a seal.

▼ **Public stand-up toilet** in France, 1886. It had a glazed surround and a syphonic flushing system.

▼ **The wash-down closet** of 1889 was the forerunner of the one we use today. It was simple and efficient but very noisy.

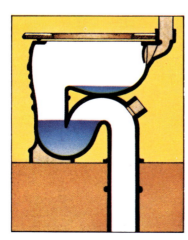
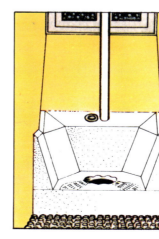
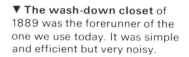

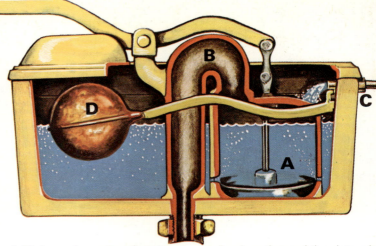

▲ **Cistern of a water closet.** When the chain is pulled, the unit (**A**) rises and forces water over the bend (**B**) at the top of the flushing pipe. This starts a syphonic action and the cistern is emptied. Water flows in again (**C**) and when the cistern is full the ball float (**D**) rises and cuts off the supply of water.

The development of the "wash-down" closet

This was invented in about 1889 by Bostel, an Englishman, and worked on the same principle as the one we use today. It was simple, it worked efficiently and the bowl was made in one piece, so there were fewer places where germs could collect.

But more was needed than efficient water closets and it was many years yet before sanitary engineers were able to control sewage arrangements safely. It took time to teach plumbers how to install the new closets and how to lay the pipes in new and complicated ways. People were horrified when outbreaks of fever in a town in southern England attacked well-to-do houses, but not poorer ones. Badly installed water-closets *inside* the house were a greater risk than simple earth closets *outside* the house, as in poorer homes.

◄ **Doulton's patent pedestal closet.** The basin was lipped to form a slopsink and urinal when the seat was raised. This closet could, if needed, be flushed by simply raising the seat. It was made about 1888.

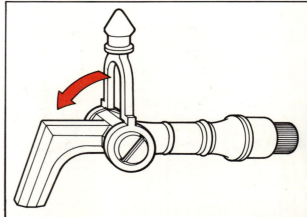

Baths, Taps, Geysers and Boilers

Baths and water taps

Most people washed themselves very rarely until they had running water and it was usual to have a bath only for medicinal reasons. Vapour baths and steam baths were used to cure a variety of ailments, and to relieve pain. Country people got their water from wells and pumps but even in the early 19th century supplies in towns were still very bad. In poor districts standpipes in the streets were turned on for only an hour a day and people often fought to get their buckets filled in the short time allowed. When they could afford it they bought water from a water-carrier but this was probably too expensive to use for washing.

Even after clean running water was available in a large house it was for a long time supplied only to the basement and the servants had to carry cans of water up and down stairs. Most baths were taken cold and were filled and emptied by hand until the 1850s. Then, wash-basins with taps and baths with overhead shower tanks began to be advertised in trade catalogues. The water for these was raised by hand pumps.

Baths of many different shapes were made from the 1850s onwards, but we can tell that they were a novelty because many advertisements gave instructions on how to wash in them! Even in 1900 the user of the bath was still being referred to as "the patient". In 1880 a "soap bath" was described. This was for cleaning, not for curing, and involved using warm or hot water and could even be taken in front of a fire. For rich families, at least, times were changing.

How a tap works. The stem of the tap is threaded inside and has a washer on the end. When the handle is turned, the washer is screwed downwards and is forced onto the seat of the valve. This stops the water flowing through. The drawing on the left is from a catalogue of 1870 and shows a tap which was opened by pulling down a handle. Another name for a water tap is a "faucet".

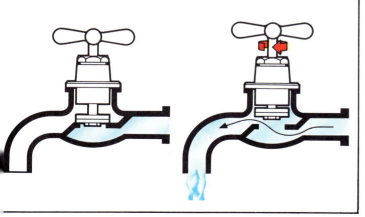

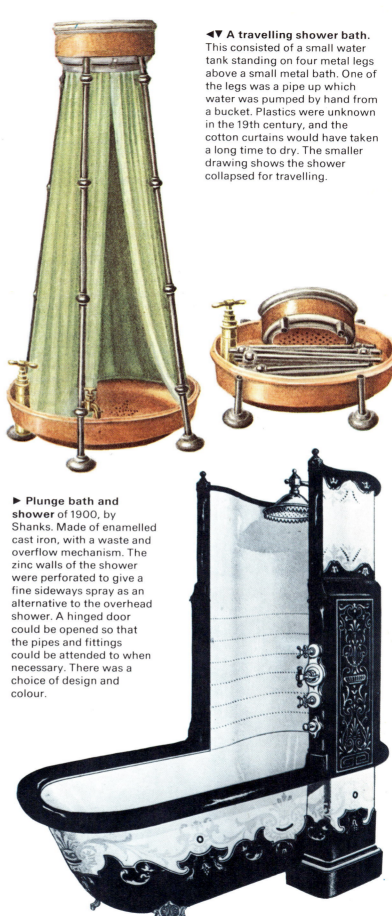

◀▼ A travelling shower bath. This consisted of a small water tank standing on four metal legs above a small metal bath. One of the legs was a pipe up which water was pumped by hand from a bucket. Plastics were unknown in the 19th century, and the cotton curtains would have taken a long time to dry. The smaller drawing shows the shower collapsed for travelling.

▶ Plunge bath and shower of 1900, by Shanks. Made of enamelled cast iron, with a waste and overflow mechanism. The zinc walls of the shower were perforated to give a fine sideways spray as an alternative to the overhead shower. A hinged door could be opened so that the pipes and fittings could be attended to when necessary. There was a choice of design and colour.

Heating water

Having a bath gradually became more acceptable and people began to adapt one room as a separate bathroom. For a long time most "bath" rooms were still converted bedrooms, but by the end of the 19th century some new homes were being designed with separate bathrooms. Many families continued to use portable baths for very much longer and folding ones were advertised, looking something like small wardrobes.

Hot water was available in the bathrooms of the well-to-do houses long before anybody thought of *piping* it upstairs. It was heated on the kitchen fire or in the boiler of a range and there were still plenty of servants to do the necessary carrying to and fro. The next stage was to carry cold water to the bath and heat it on the spot. An article in *The Magazine of Science* in 1842 stated that "many copper and tin baths have lately been constructed in London, with a little furnace attached to one end" and a bath of that type was shown at the Great Exhibition of 1851.

Hot water boilers built into kitchen ranges were still popular but they were dangerous. Bad explosions often occurred in large houses, when preparations were being made for an evening party. There would be a large demand for hot water from all parts of the house, so the boiler would be at full heat. But if the water was drained off too quickly, leaving the boiler empty, the compressed steam would cause an explosion.

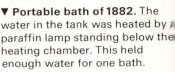

▲ **Gas bath of 1882,** with a towel warmer. The gas burner swung out for lighting and it was important to remember to turn it off before sitting down in the bath!

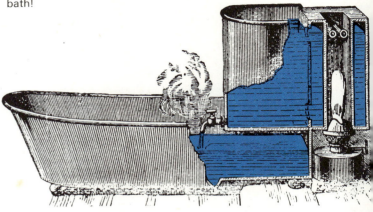

▼ **Portable bath of 1882.** The water in the tank was heated by a paraffin lamp standing below the heating chamber. This held enough water for one bath.

▼ **Maughan's gas geyser of 1868** was the first heater to provide hot water whenever it was needed, without having to carry water by hand.

▼ **Hammond's gas water heater of 1900.** The water passed up and then down the cylinder through a coiled pipe. It had a bunsen-type burner.

▼ **How an instantaneous gas geyser works.** When the cold tap (**A**) is turned on, water flowing in causes the gas to flow to the burner (**B**) where it is lit by a pilot light (**C**). The flames from the burner heat the water in the pipe (**D**), and the hot water flows out (**E**). There is no storage of water.

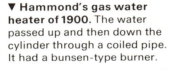

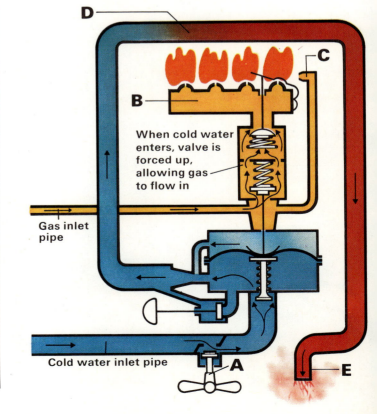

Gas inlet pipe

When cold water enters, valve is forced up, allowing gas to flow in

Cold water inlet pipe

Geysers

The gas geyser, invented by Benjamin Maughan in 1868, heated a running supply of water. A geyser contains a water-pipe of copper or iron, which passes through a metal cylinder. It is heated by gas burners in direct contact with the pipe. Early geysers were dangerous, as they were not fitted with outlets for the gas fumes and there were frequent explosions if people did not follow the complicated instructions.

Those who used geysers had to have one at every basin or bath. It was a great improvement when multi-pressure geysers were invented, for one of them could feed every tap in the house. But there was a snag in this: if two people wanted to draw hot water at the same time one would find the water cold. Modern geysers have temperature controls which can be set for summer or winter use.

Heating water by electricity

Until 1945 no-one could find a safe way of transferring heat from an electrically-heated wire to water. Nowadays, however, sealed immersion heaters are widely used. Their heat is controlled by a thermostat, and they are completely insulated to maintain the temperature.

Solar energy

The sun's energy can be used for water heating in many countries. Large black metal plates, called "receivers", are built onto rooftops and insulated tanks under them store hot water which is piped into the house.

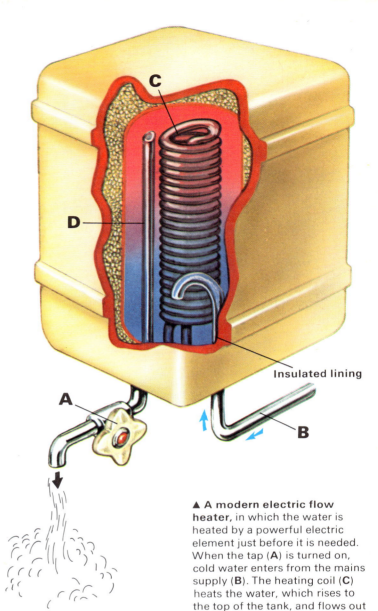

Insulated lining

▲ **A modern electric flow heater,** in which the water is heated by a powerful electric element just before it is needed. When the tap (**A**) is turned on, cold water enters from the mains supply (**B**). The heating coil (**C**) heats the water, which rises to the top of the tank, and flows out through the hot water outlet (**D**).

▼ **Ewart's "Califont" multi-point pressure geyser of 1899.** Installed in the basement, it provided "hot water instantly night or day at every tap in the house" according to this advertisement. A later improvement was a gas-heated water storage tank.

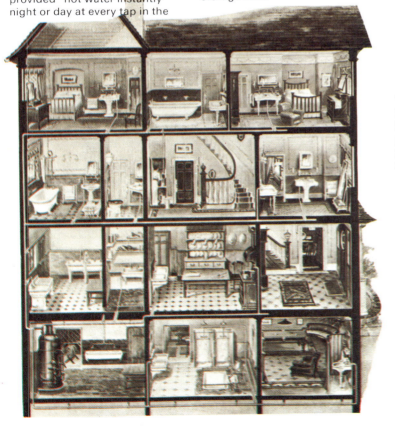

▼ **Advertisement of 1914 for Fletcher's "Ajax" water heater,** illustrating the advantages of their gas geyser, and emphasising the pleasure of "a hot bath at any moment". These geysers could be installed in any room in the house, where hot water was needed. When the new geysers appeared many "bathrooms" were created by converting spare bedrooms.

Vacuum Cleaners

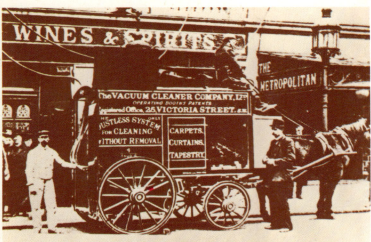

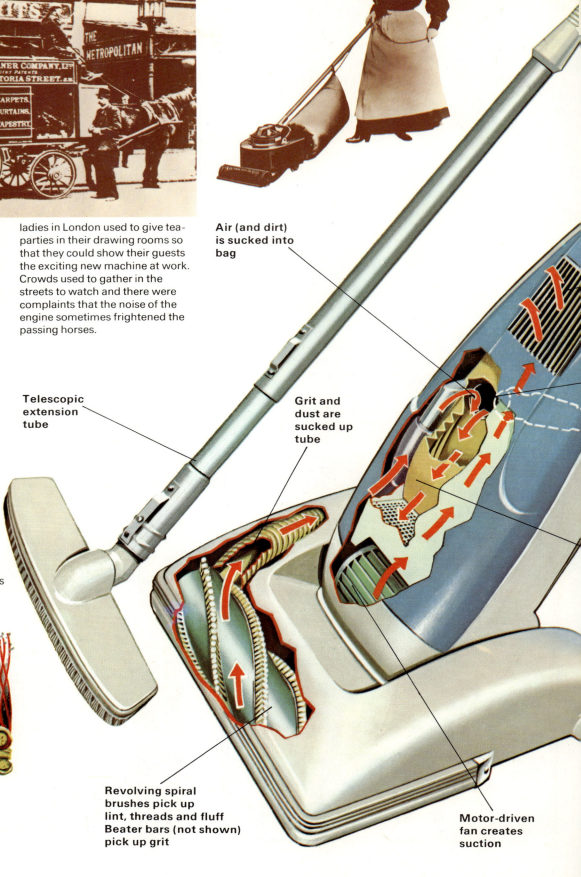

► **Early Hoover Cleaner,** patented in 1908. It had an electric motor, a foot-operated switch and a fan unit on small wheels.

▲ **The Vacuum Cleaner Company's van** at work in a London street in 1903. The horse-drawn van contained a strong pair of bellows worked by a petrol engine. A hose was put through the window of the room to be cleaned. This process cost customers a lot of money. Many ladies in London used to give tea-parties in their drawing rooms so that they could show their guests the exciting new machine at work. Crowds used to gather in the streets to watch and there were complaints that the noise of the engine sometimes frightened the passing horses.

► **A modern vacuum cleaner** has a powerful electric motor which revolves beater bars and spiral brushes which pick up dirt by suction. This model can also be used with a hose.

▼ **Cross-section of a soiled carpet.** Modern vacuum cleaners can remove all types of dirt and dust at different levels.

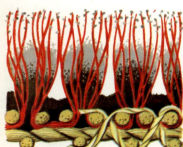

 Surface litter

Surface dust

Embedded dirt and grit

Air (and dirt) is sucked into bag

Grit and dust are sucked up tube

Telescopic extension tube

Revolving spiral brushes pick up lint, threads and fluff Beater bars (not shown) pick up grit

Motor-driven fan creates suction

Stretchable
hose

On-off switch

Electric flex
to power point

After main dirt has
been caught in bag,
exhaust air is
cleaned by filters

Nozzle fits into back
of cleaner, to convert
it for use with a hose

Disposable
bag catches
dirt

The history of the vacuum cleaner

Small domestic vacuum cleaners were first made in about 1904 by an Englishman named H. Cecil Booth. At first they needed two people to work them – one to turn the handle working a pair of bellows and the other to guide a nozzle to suck dust and dirt from the area to be cleaned.

Booth's cleaner was developed from his patent of 1901 for a device which drew air through a cloth which caught the dust. He founded the British Vacuum Cleaner Company which used large pumps drawn by horses (*see opposite*). There was great enthusiasm for the new "suction machines" and soon other small domestic models were made which were powered by electricity and needed only one operator.

Many servants refused to use vacuum cleaners at first, some because they were frightened of electricity and others because they feared they might lose their jobs if all the cleaning was to be done by machinery. But after World War One, the shortage of servants led more and more families to buy an "electric suction sweeper"

Latest developments

Modern vacuum cleaners have more powerful suction than earlier ones; they are lighter to use and more hygienic. In new houses an electric pump can be installed centrally with pipes running from each room, along which the dust can be conveyed to a large, removable container.

EARLY VACUUM CLEANERS

▲ A patent drawing of **1908** of an early electrically-driven cleaner.

▲ "Baby Daisy" cleaner, **1910**. The large bellows were hand-operated.

▲ The "Wizard", **1912**, moved on a trolley. A wheel worked two bellows to pump the air.

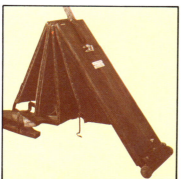

▲ The "Cyclone", **1914**. The bellows worked by pressing and releasing a foot-rest at the back.

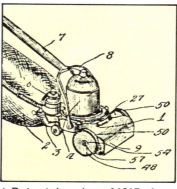

▲ Patent drawing of **1915** of an American electrically-operated vacuum cleaner.

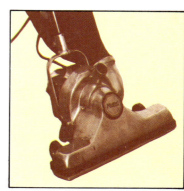

▲ "Magnet" electric cleaner, **1917**. A belt drove an agitator or roller covered with brushes.

Washing Machines

◄ **Tindell's Scotch mangle** of 1850 looked very like a printing press. Washing was laid on a linen-covered shelf and wound round three rollers. Underneath was a press and two drawers.

► **An early wringer,** made in the United States in 1847. It had no rollers; the motion of hand-wringing was imitated by twisting a sack with clothes in it.

▲ **Box mangle of 1860,** used in most big houses. The clothes were wrapped round wooden rollers and put under a box filled with heavy stones. A wheel moved the box to and fro.

▲ **Washer and spin dryer,** made in 1924 by Savage of New York, but not used in Europe till 40 years later. After the washing a vertical spindle turned at high speed and dried the clothes.

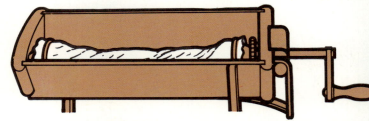

▼ **Morton's patent steam washing machine** of 1884 had to be turned by hand. Notice that the stand is very similar to those of sewing machines of the time.

The development of the washing machine

Washing has always been done by rubbing clothes in water by hand, but from the middle of the 19th century hand-operated wooden washing machines were made. The clothes were put into a wooden box and tumbled over by turning a handle.

An advertisement in *The Poor Man's Guardian,* a British newspaper of 1832, referred to a washing machine "which supercedes the necessity of hiring charwomen, makes the Linen a far better colour without half the injury or wear, and in one-third of the time; they are turned in a manner similar to a mangle, and will wash from 12 to 32 shirts in 30 minutes, and the work may be chiefly done by a boy".

Washing machines with electric motors were made in 1914. At first they were not insulated, so that water often dripped onto them and gave an electric shock to the user. In the 1920s a new kind of washing machine was designed; the tub was made of tinned copper, the motor was totally enclosed for safety and drove both the "dolly", which rotated the clothes in the water, and a mangle.

The first mangles were box mangles, invented in the early 19th century; but as early as 1677, the scientist Robert Hooke noted the invention of a device for wringing clothes, rather like the 1847 model shown above. Later the now familiar type was introduced, with horizontal rollers rotated at first by a handle and later by an electric motor.

The domestic spin dryer was first introduced in America in 1924 but only became popular after World War Two.

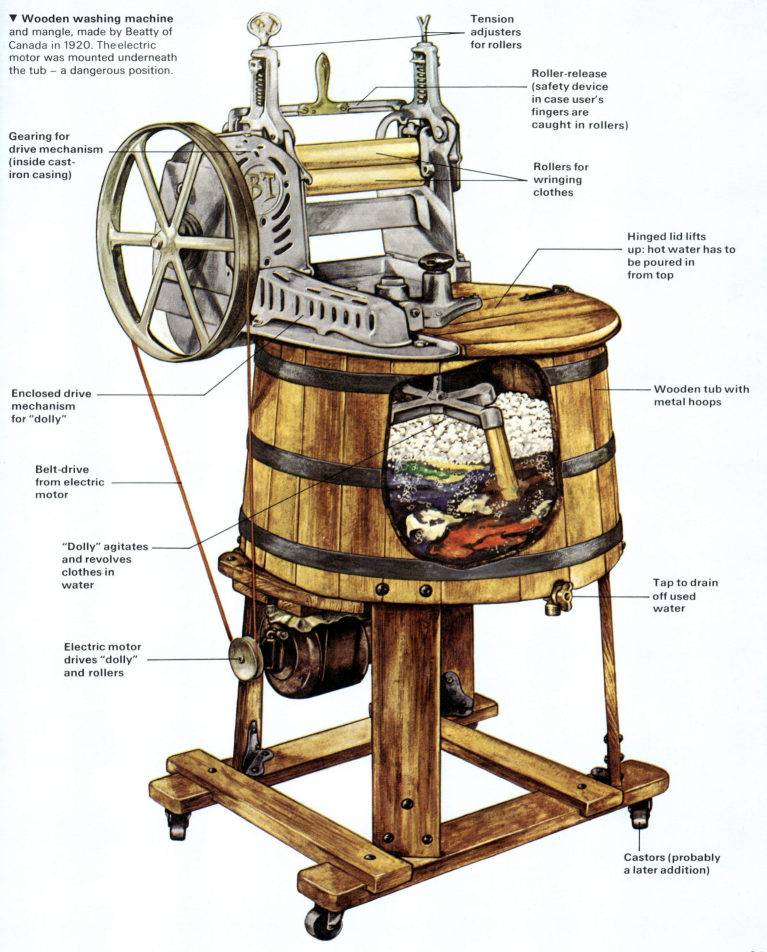

▼ **Wooden washing machine** and mangle, made by Beatty of Canada in 1920. The electric motor was mounted underneath the tub – a dangerous position.

Tension adjusters for rollers

Roller-release (safety device in case user's fingers are caught in rollers)

Rollers for wringing clothes

Gearing for drive mechanism (inside cast-iron casing)

Hinged lid lifts up: hot water has to be poured in from top

Wooden tub with metal hoops

Enclosed drive mechanism for "dolly"

Belt-drive from electric motor

Tap to drain off used water

"Dolly" agitates and revolves clothes in water

Electric motor drives "dolly" and rollers

Castors (probably a later addition)

Zips, Locks, Razors, Thermos Flasks

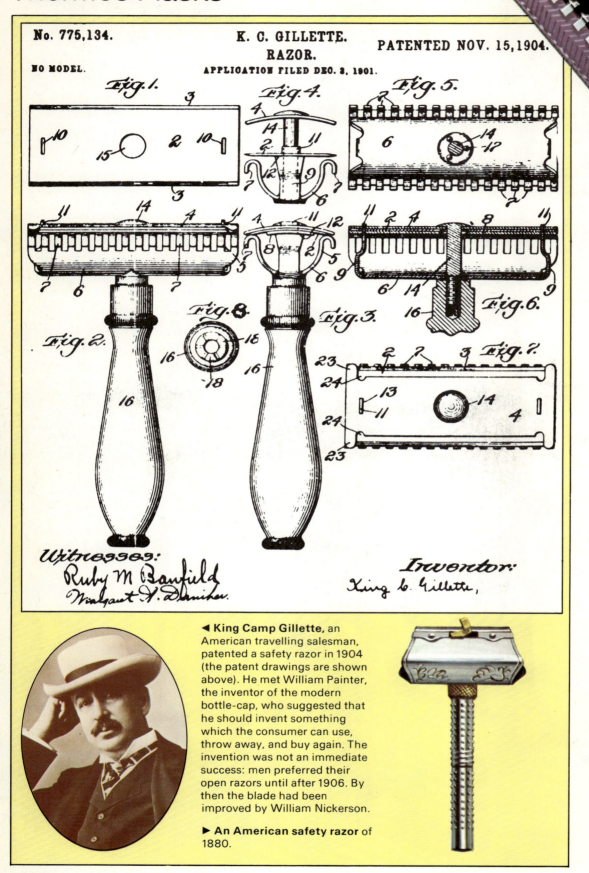

No. 775,134.

K. C. GILLETTE.
RAZOR.
APPLICATION FILED DEC. 3, 1901.

PATENTED NOV. 15, 1904.

NO MODEL.

Fig. 1. Fig. 4. Fig. 5. Fig. 2. Fig. 8. Fig. 3. Fig. 6. Fig. 7.

Witnesses:
Ruby M Banfield
Margaret A. Danaher.

Inventor:
King C. Gillette,

◄ **King Camp Gillette,** an American travelling salesman, patented a safety razor in 1904 (the patent drawings are shown above). He met William Painter, the inventor of the modern bottle-cap, who suggested that he should invent something which the consumer can use, throw away, and buy again. The invention was not an immediate success: men preferred their open razors until after 1906. By then the blade had been improved by William Nickerson.

► **An American safety razor** of 1880.

Zips

In the 19th century inventors were trying to find a better way of doing up garments than with a great many hooks, eyes, buttons or laces. Dr. Sundback, a Swedish American, patented a slide-fastener in 1914. In 1931 the patents expired and since then "zips" have become all-purpose; they are made all over the world and are used for many different purposes.

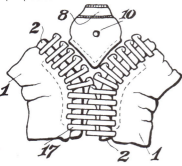

▲ **A zip fastener** consists of two chains of metal or plastic teeth attached to a strip of fabric. Each chain has many teeth, each one with a small projection which fits into a recess in the opposite chain.

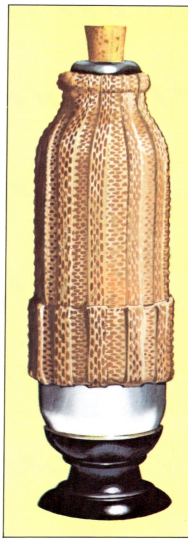

The Thermos flask

"Thermos" means hot. Scientists had known since the 17th century that heat cannot pass through a vacuum, but it was not until 1892 that Sir James Dyer, an Englishman, made a vacuum flask to keep hot liquids hot and cold liquids cold.

The glass bottle in a vacuum flask has double walls, and the air between them is removed, creating a vacuum. The inner surfaces of the two walls are silver coated: with hot contents, inner heat is radiated back into the bottle, and with cold contents external heat is radiated away from the bottle.

In 1904, a prize was given for the best name for a vacuum flask, and "thermos" was the winning entry.

1. Cap/Beaker
2. Stopper
3. Outer casing
4. Vacuum bottle (twin layer of glass)
5. Hot liquid
6. Flexible nylon support for glass bottle
7. Nipped off end of vacuum tube

◀ Sir James Dyer made a vacuum flask for his baby son's milk but his mother-in-law doubted its efficiency and knitted a woollen "cosy" for it.

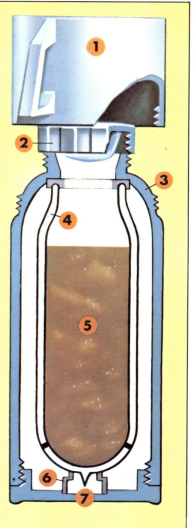

Ancient and modern locks

The ancient Egyptian lock on the left is very much like a modern Yale lock. The key's prongs lift a set of rods inside the lock so that the wooden bolt can be slid to the right. In the locked position the rods prevent the bolt from moving.

In the Yale lock (invented by Linus Yale in 1848) the key raises a set of metal pin-tumblers to the correct height, freeing the barrel.

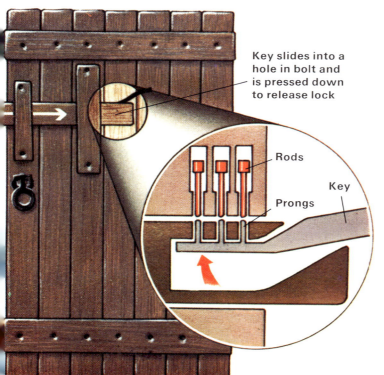

Key slides into a hole in bolt and is pressed down to release lock

Rods

Prongs

Key

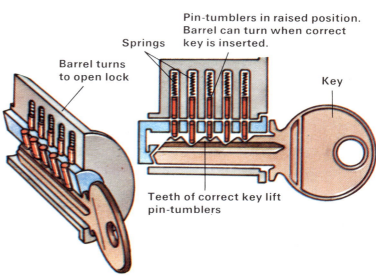

Pin-tumblers in raised position. Barrel can turn when correct key is inserted.

Springs

Barrel turns to open lock

Key

Teeth of correct key lift pin-tumblers

Sewing Machines

▼ **Lock-stitch machine** with an oscillating hook, introduced by Singer in 1879. The bobbin lay horizontally, and not vertically as in most modern machines. This drawing is cut-away to show the cranks, levers and cams which made the movements quieter. The stand for the treadle is not shown.

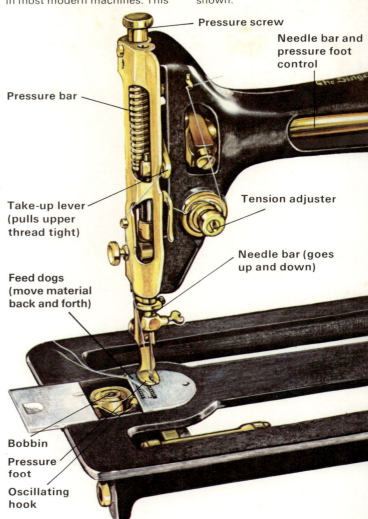

Pressure screw

Needle bar and pressure foot control

Pressure bar

Take-up lever (pulls upper thread tight)

Tension adjuster

Needle bar (goes up and down)

Feed dogs (move material back and forth)

Bobbin

Pressure foot

Oscillating hook

▲ **The best-known name in the world of sewing machines is Singer.** Isaac Singer was a German who went to America and was a farmer, machinist and actor before he invented a sewing machine in 1850 for use in garment factories. Since then Singer have become the biggest sewing machine company in the world. They make 250 kinds of machines, own their own forests and saw-mills and produce their own electric motors. Hire-purchase was started by Singers when they made the first domestic sewing machine in 1856.

▼ **How a stitch is made in the flying shuttle machine** (usually manually operated). The lower thread is carried on the bobbin inside a shuttle. The threaded needle goes through the fabric, carrying the thread **(1)**. As the needle rises, it leaves a loop of thread under the fabric. The shuttle goes through this loop **(2)**, pulling the under thread behind it to form a stitch **(3)**. The fabric moves forward; the shuttle returns and the threads are pulled tight **(4)**.

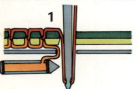
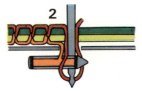
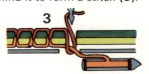
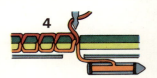

▼ **How a stitch is made in a machine with an oscillating hook.** In this type of machine (usually electrically driven) the shuttle moves around a central bobbin, which remains stationary. **1.** As the needle descends, the shuttle turns. **2.** When the needle rises, it leaves a loop which is caught by the hook on the shuttle. **3.** It pulls the upper thread over the bobbin to link it with the lower thread. **4.** The loop slips off as the take-up lever pulls the thread tight **(5).**

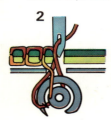
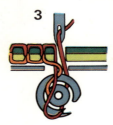

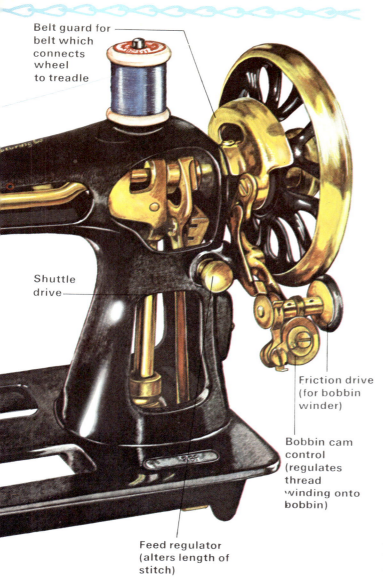

Belt guard for belt which connects wheel to treadle

Shuttle drive

Friction drive (for bobbin winder)

Bobbin cam control (regulates thread winding onto bobbin)

Feed regulator (alters length of stitch)

The development of the sewing machine

From the late 18th century onwards, many men tried to invent a machine which would sew, in order to save working time in workshops and factories.

The earliest sewing machine was made in 1790 by a London cabinet-maker, Thomas Saint, but his patent remained undiscovered for 84 years. In 1810 a German hosiery worker, B. Krems, made a chain-stitch machine, but this too was unsuccessful. Then in 1830, a French tailor, Thimonnier, made a machine which was successful enough to attract a great deal of opposition from workers in various industries. The machines were used to make army uniforms, but the workers broke up the machines because they were afraid they might lose their jobs.

In the 1840s and 1850s the sewing machine industry started successfully in the United States, but many men had similar ideas at the same time and there were legal battles about who had the patent rights on various parts of the machinery. This "sewing machine war" ended in 1856 when several companies pooled their ideas.

The effects of the invention

From then onwards the sewing machine had an enormous effect upon people's lives. By 1900, 20 million machines were being produced. Manufacturers could profitably make ready-to-wear clothes for the first time. Although wealthy men and women still liked to have their garments made by hand for some time afterwards, most families benefitted greatly from the new, cheaper, factory-made clothing.

The sewing machine was the first complex machine, after the clock, to be used in the home. It saved women an enormous amount of time and labour in making the family clothes, particularly after the introduction of the first portable electric machine in 1921. A domestic sewing machine can now do 1,500 stitches a minute!

▼ **Pioneer machines.** In 1846 Howe made this lock-stitch machine in America. It had a curved needle attached to a swinging lever. The cloth was attached vertically to a strip of metal and moved along with it. The first Singer machine in 1851 was for factory use. It was packed in a box which formed a stand with a treadle in it. The Wheeler and Wilson machine of 1854 was quieter because it had no moving shuttle. The needle thread was caught by a rotating hook inside a bobbin below. Notice the elaborate decoration in gold paint and inlaid mother-of-pearl.

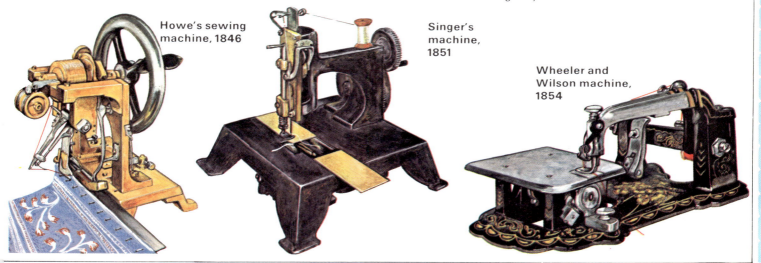

Howe's sewing machine, 1846

Singer's machine, 1851

Wheeler and Wilson machine, 1854

Typewriters

▲ **Lillian Sholes,** daughter of the inventor, demonstrating her father's new writing machine, which had no shift key and printed only in capital letters. Christopher Latham Sholes,

an American printer, worked for years with two colleagues, trying to make a writing machine. In 1878 he patented a "type-writer".

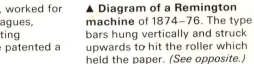

▲ **Diagram of a Remington machine** of 1874–76. The type bars hung vertically and struck upwards to hit the roller which held the paper. *(See opposite.)*

▼ **A modern manual typewriter** works by levers which, when a key is pressed, cause the corresponding type bar to strike the paper above, through an inked ribbon.

The first writing machine

The first mention of a machine which would "write" was in 1714. Anne, Queen of England, granted a patent in that year to an engineer, Henry Mill, who had invented a "method for the impressing of letters one after the other". The Mills machine was not made for sale and we do not know what it looked like. In those days it was not necessary to make a drawing or model of a new invention before obtaining a patent, as it is now.

During the next hundred years, many people tried to invent writing machines for blind people to use, but it was not until 1829 that an American, William Burtt, was granted a patent for "Burtt's Family Letter Press". Various types of machines were produced in the 19th century, few of which resemble the familiar modern machines. The aim was to produce a machine that would write as fast as one could write with a pen.

Sholes and Remington

The first commercially successful typewriter was made by Christopher Sholes, an American, and his colleagues, Glidden and Soule. It developed from Shole's work on a machine suitable for numbering book pages consecutively. In 1873 the Remington Fire Arms Company became interested in the new machine. Now that the American Civil War was over and guns were not wanted in such great numbers, this company was using its machinery to make sewing machines. After 1874 they made typewriters as well.

At first they lent them to American companies, but gradually they were found to be so useful that people started to buy them. More than 300 different makes of typewriter have been made since then but the layout of the keyboard has remained the same.

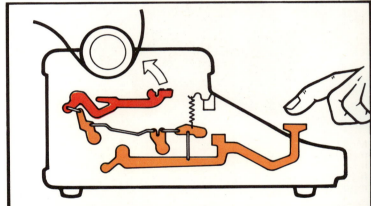

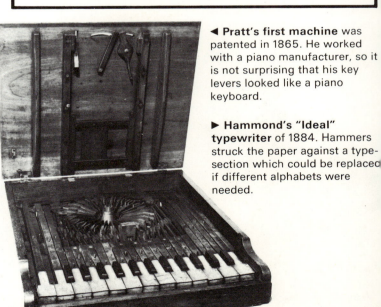

◄ **Pratt's first machine** was patented in 1865. He worked with a piano manufacturer, so it is not surprising that his key levers looked like a piano keyboard.

► **Hammond's "Ideal" typewriter** of 1884. Hammers struck the paper against a type-section which could be replaced if different alphabets were needed.

► **The Sholes-Glidden machine** made by Remington in 1876 wrote in capital letters only. The arrangement of the keyboard has since been adopted for nearly all typewriters.

Carriage had to be folded down for typing, and lifted up (as here) so that what had been typed could be read

Typewriter ribbon

Type-bars hung down in a cylindrical "basket"

Rope pulled carriage to right

Spool for ribbon (another one on opposite side)

Metal cover (removed to show inner workings) fitted on front of machine

Carriage return handle

Space-bar (for spaces between words)

► **The "Mignon" typewriter** was made in Germany in 1904. A pointer was used to select the letter. This positioned the letter on a cylindrical typehead, ready for typing.

▲ **The IBM "golfball" typehead.** The typist can change the style of type on a machine simply by changing one golfball-shaped typehead for another one.

37

The Telephone

Until the 19th century, if you wanted to speak to someone you had to go *to* them. By 1836 an electric telegraph had been invented, on which messages could be sent in code, but it was not until after a Scotsman, Alexander Graham Bell, had demonstrated his telephone in 1876 that people could speak to each other over a great distance.

It was in Canada that the young Bell became interested in electricity and began to consider how words might be turned into electrical impulses and transmitted along a wire. Much previous experimentation had been done on telephones by other people, including a German, (Philip Reis), an Italian, (Meucci), and, at the same time as Bell, an American, (Elisha Gray).

Bell's first telephone consisted of a thin sheet of metal (called a diaphragm) suspended in front of an electro-magnet. When the sound waves (caused by his voice) struck the diaphragm, electricity was generated in the coils of the electro-magnet. These electric currents were transmitted to a telephone in another room along a wire. When they passed through coils of a similar electro-magnet in the receiver they caused the receiver diaphragm to vibrate and create the original sound. His assistant, Watson, heard Bell shouting "Mr. Watson, come here! I want you".

Many other men improved the telephone after Bell. The first conversation between England and Europe took place in 1891 and by 1923 it was possible to speak from London to New York. Now most exchanges are fully automatic and subscribers can dial numbers in many countries in the world. Some calls over long distances are now carried by the space satellite Telstar.

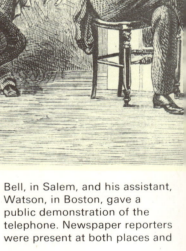

▲ **The first demonstration** of how the human voice could be heard over a long distance by means of a transmitter and a receiver. On March 14, 1877, Bell, in Salem, and his assistant, Watson, in Boston, gave a public demonstration of the telephone. Newspaper reporters were present at both places and

▶ **A wall telephone of 1900.** The electrical power required to signal the operator and for transmitting was obtained from a battery at the exchange.

▼ **Bell's telephone,** in 1875, was the first instrument to transmit sounds electrically. By this "gallows frame" telephone, faint sounds could be heard, but it was not possible to transmit actual words until the following year, when Bell had adjusted his instrument so that it "talked". The first telephone call took place in Bell's house.

▼ **The Gower-Bell wall telephone** of 1880 was for many years supplied by the British Post Office. The transmitter was in the lid of the wooden box and the receiver was connected to ear pieces by two speaking tubes.

▼ **Table telephone of 1885.** The company who made it had to stop manufacture in 1888 because they were accused of copying the patents of Bell and Edison. The receiver hung on a hook and the caller spoke into the microphone.

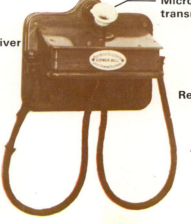

Receiver

Microphone/ transmitter

Receiver

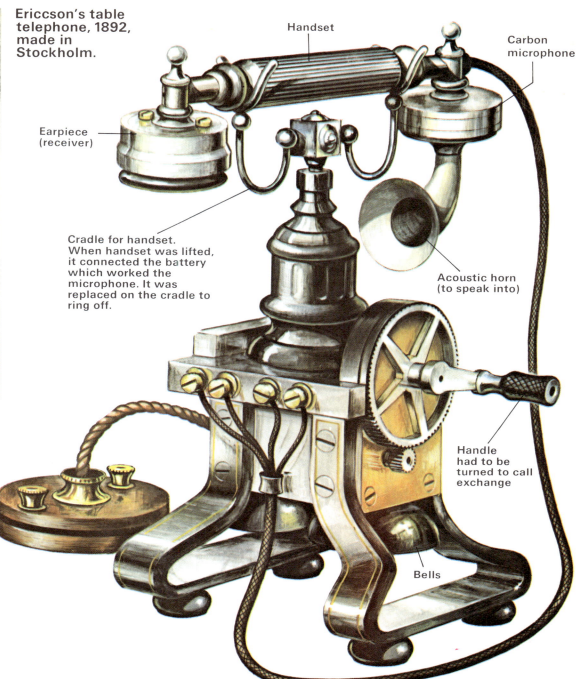

Ericcson's table telephone, 1892, made in Stockholm.

Handset

Carbon microphone

Earpiece (receiver)

Cradle for handset. When handset was lifted, it connected the battery which worked the microphone. It was replaced on the cradle to ring off.

Acoustic horn (to speak into)

Handle had to be turned to call exchange

Bells

compared notes afterwards to make sure that there had been no cheating.

► Table telephone of 1890, made by Ericcson of Sweden until the 1920s. The wet battery (not shown) that worked the microphone was in the subscriber's house as part of the installation. To make a call, the subscriber lifted the handset (thus connecting the battery) and turned the handle. This generated an electric current which called up the exchange.

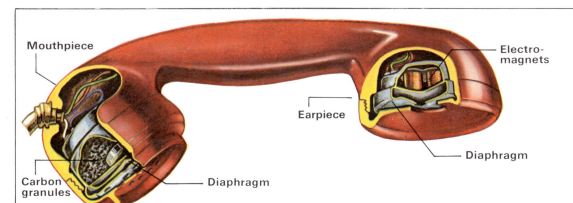

How a telephone works

The speaker's voice sets up vibrations in the air. These vibrations (sound waves) strike a steel diaphragm. The diaphragm vibrates against carbon granules in the microphone and this sets up an electric current. At the other end of the line, the process is reversed. An electromagnet is affected by the electric current and causes another diaphragm to vibrate.

Mouthpiece

Electro-magnets

Earpiece

Diaphragm

Diaphragm

Carbon granules

The Gramophone

The invention of the phonograph

The first "talking machine" was made in the United States in 1877 by Thomas Edison. He had already made an instrument that would convert into sounds the dots and dashes of a message telegraphed in Morse Code, and imprint them on a piece of oiled paper. This "record" could be used to repeat the message later. Later that year Edison developed some of Alexander Bell's ideas and made a machine that would reproduce the human voice.

This was the "phonograph" which consisted of a drum with a recorder and a reproducer on either side. To record, the speaker turned the handle to revolve the drum and spoke into the mouthpiece. At the end of the mouthpiece was a diaphragm, in the centre of which was a stylus. This pressed on a tinfoil plate on the drum (as the drum revolved) and made scratches which corresponded to the sound vibrations of speech. To reproduce the message, the drum was rotated backwards, and the process reversed. The machine gave only a distorted reproduction of speech; but improvements soon followed.

The graphophone

Alexander Bell became interested in the phonograph and joined Edison to carry out further experiments. Together they produced a "graphophone" in 1885, in which the sound impressions were recorded on the wax-coated surface of a cardboard cylinder. They experimented with many kinds of cylinders, discs and tapes.

▲ **Listening to a message from America.** This engraving of 1888 shows a family using Edison's phonograph, with hearing tubes. Later models had a horn attached.

▼ **A clockwork toy talking machine of 1903.** There was evidently still much experimentation going on because one French scientific magazine suggested that discs for this machine could easily be made out of chocolate!

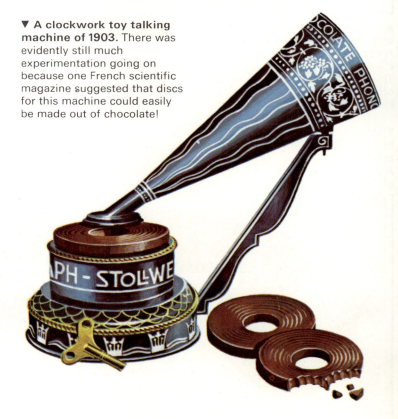

▲ **Edison's tinfoil phonograph of 1877,** the first machine to record and reproduce sounds. The handle was turned one way to record, and the opposite way to play back the sounds. The idea was not new, however. Earlier that year, a Frenchman, Charles Clos, had put forward a brilliant idea for a phonograph, but had failed to make a working model.

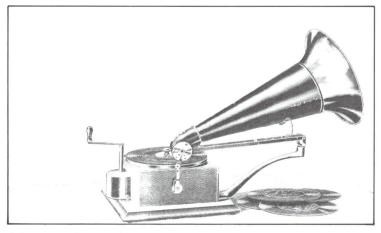

▲ **A gramophone using discs was available by 1900.** A small handle turned a clockwork motor inside the box. The 7 in. (175 mm.) discs were of vulcanite – a mixture of india-rubber and sulphur, hardened by intense heat. The greatest operatic stars – Chaliapin, Caruso, Melba and Patti – recorded on these discs, which were very expensive to buy.

▼ **In 1910 Captain Scott and his companions** took one of the new gramophones on their Antarctic expedition. Here one of their huskies appears to be listening intently. Its pose looks very similar to that of the dog in the symbol used by the "His Master's Voice" company in Britain.

HOW A STEREO RECORD PLAYER WORKS

◄ **The wavy-sided spiral groove in a stereo record** corresponds to the pitch and volume of the original sound. The vibrations set up in the stylus moving along the groove are converted into electrical signals. These are passed by wire to the amplifier unit.

◄ **The amplifier** boosts the signals from the stylus until there is sufficient power to "drive" the loudspeakers. A stereo unit has two amplifiers to produce two boosted signals – one for each loudspeaker.

◄ **Loudspeakers.** Power from the amplifier is used to vibrate the cone in the loudspeaker (made from paper or plastic). This in turn produces sound waves in the air.

▼ **Stylus size** has decreased since the sapphire used on early 78 r.p.m. records. A small stylus with a lightweight mounting reduces wear and sound distortion.

Stylus for
◄ 78 rpm
► 33/45 rpm

The development of the gramophone

Emile Berliner, a German who emigrated to the United States, demonstrated a gramophone in 1888 and in 1889 a gramophone first appeared for sale in Germany, curiously enough being advertised as a children's toy. By 1893 there were both hand- and electrically-driven models on sale in the United States. They played hard rubber discs 5 in. (125 mm.) across, called "plates", on which the track was etched chemically. Their great advantage was that several duplicated recordings could be made from one master. The pre-recorded cylinders which came with the phonograph were more expensive because they had to be recorded individually.

The quality of reproduction of these early "talking machines" was very poor, but selling them was a luxury trade. This prospered only in the winter, so a dealer often sold bicycles as well, to keep his business going in the summer. Larger discs were available by 1903 and the first disc to be recorded on both sides was the "Odeon", made in Germany in 1904. Some machines used blank cylinders, mainly for office dictating. Until 1929 models for home-recording were available, supplied with a device for shaving off one imprint before adding another – a very early ancestor of our modern tape-recorder. The first slow-speed turntable was produced in 1948.

Radio

Radio waves were detected in 1888 by Heinrich Hertz in Germany. He discovered that when he passed an electric current through a metal coil ending in two knobs, a spark jumped across the air gap between them. At the same time, an identical spark jumped across the space between two knobs on a metal ring set up a few feet away from the coil, *although there was no visible connection between the coil and the metal ring* The electro-magnetic "Hertzian" waves which he had discovered moved with the speed of light.

Radio works by converting the electrical signals from a radio transmitter into an electro-magnetic ("Hertzian") wave, which spreads out from it. A receiving aerial traps the wave and converts it back into electrical signals which can be boosted and turned into audible sounds by a receiving radio set.

In 1894 Sir Oliver Lodge in England and Professor Branly in France transmitted and received the first radio message, by the dots and dashes of the Morse code. This code was already widely used to send messages along a wire by telegraphy and early radio was called "wireless telegraphy"

Professor Popov, in Russia, succeeded in receiving messages over longer distances by using an aerial and his ideas were followed by Marconi, who managed to make an electric bell ring by wireless transmission, and sent a message in Morse across a distance of five miles. The transmission of speech had to await the invention of the valve by Sir John Fleming in England and Lee de Forest in the United States.

By 1913 it was possible to transmit voices by radio over considerable distances. Regular broadcasting started in the United States in 1920, and in Britain two years later.

▲ **Guglielmo Marconi** (1874-1937) an Italian electrician, steadily increased the distance over which wireless signals could be sent and in 1901 they were heard across the Atlantic. He formed the Marconi Wireless Telegraph Company in 1908 and did a great deal of work installing radios in ships.

▲ **The "radiola 17",** the first mains radio, made by the Radio Corporation of America in 1925. The receiver contained six valves and a rectifier to convert alternating current into direct current. The loudspeaker was a separate unit on top of the set. It was highly decorated.

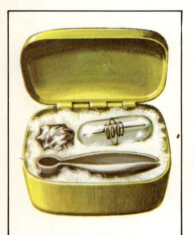

▲ **The crystal and "cat's whisker" wire** for a crystal radio set, 1925. This was the first popular radio, and it worked with headphones.

▲ **Radio cabinet of 1932.** Cabinets were highly decorated until the mid-1930s, when simple, well-designed ones were made.

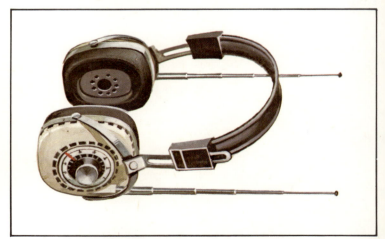

▲ **A modern Japanese radio set,** self-contained, with radio receivers, stereo headphones and aerials, all in one single unit. The Japanese electronics industry is advanced in the technique of miniaturization – making very small radios and other electrical appliances.

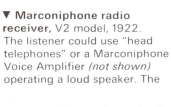

▼ **Marconiphone radio receiver**, V2 model, 1922. The listener could use "head telephones" or a Marconiphone Voice Amplifier *(not shown)* operating a loud speaker. The radio had two batteries, an internal one to drive the current through the valves, and an external battery *(not shown)* to heat up the valves. These radios had an aerial, about 100 ft. (30 m.) long. It was usually a long horizontal wire connected to a tree or a mast about 70 ft. (21 m.) from the house. There was no volume control.

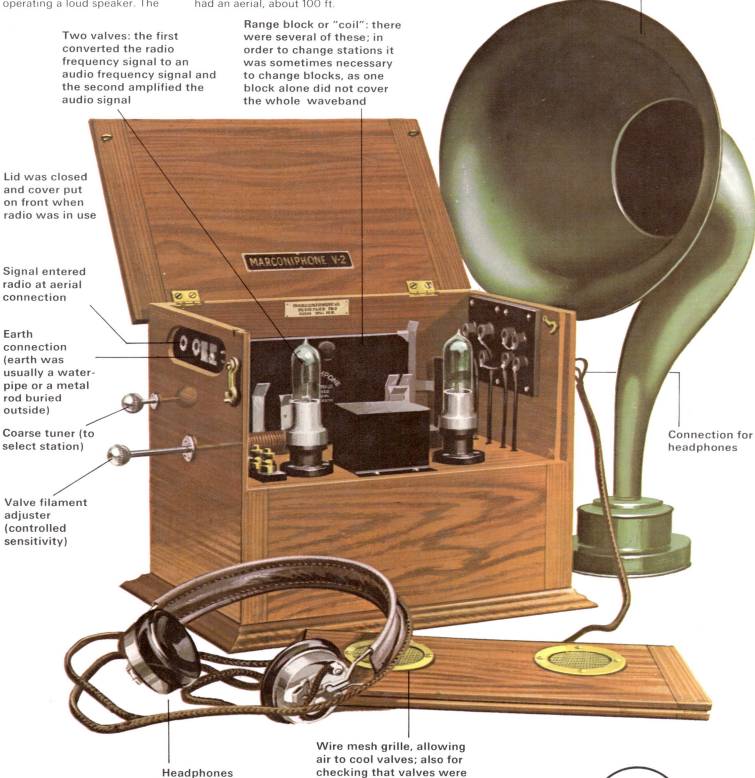

Two valves: the first converted the radio frequency signal to an audio frequency signal and the second amplified the audio signal

Range block or "coil": there were several of these; in order to change stations it was sometimes necessary to change blocks, as one block alone did not cover the whole waveband

Lid was closed and cover put on front when radio was in use

Signal entered radio at aerial connection

Earth connection (earth was usually a water-pipe or a metal rod buried outside)

Coarse tuner (to select station)

Valve filament adjuster (controlled sensitivity)

Loudspeaker (for use with Voice Amplifier)

Connection for headphones

Headphones

Wire mesh grille, allowing air to cool valves; also for checking that valves were glowing properly

► **Transistors** gradually took the place of valves after 1948. Here is one of the first, actual size.

MARCONIPHONE V-2

43

Television

▲ **Baird's 30-line Televisor** of 1930 gave small, flickering images.

▲ **Baird's first camera, 1924.** For many years the ability to "see at a distance" was sought by inventors. The first to achieve success – although the picture was very crude – was John Logie Baird. He used the odd-looking machine above fitted with lenses from a bicycle lamp! This system was improved and was broadcast for some years although the pictures were very indistinct and shaky. But his work started other people working on TV.

▶ **1930s Marconi–EMI receiver.** A big improvement came when the mechanical system was replaced by electronics using a cathode-ray tube in the receiver and electronic TV cameras.

◀▲ **Two early sets** for 405-line television, one with a mirror.

HOW COLOUR TELEVISION WORKS

Colour TV programmes can also be seen in black-and-white on a non-colour set. The colour of each point of the picture is "coded" in terms of red, blue and green by the camera. The picture is sent out as a series of electrical signals **(1)**. These signals are sent from the studios to the transmitters **(2)** which have tall masts and are often on hill-tops so that the television signal will cover a large area. As the signals spread out they become weak and so each receiver usually needs a good aerial **(3)** on the roof. The television signals travel along a feeder cable from the aerial to the receiver. Here they are amplified and then separated out into different functions: the sound signal goes to the loudspeaker;

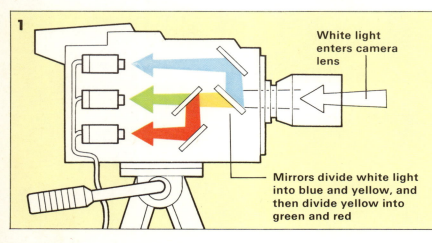

1

White light enters camera lens

Mirrors divide white light into blue and yellow, and then divide yellow into green and red

A colour TV camera is really three cameras in one. Each point of the picture is analyzed in terms of red, green and blue using three pick-up tubes that change the light values into electrical signals. These signals are then combined in an "encoder" before going to the transmitter.

2

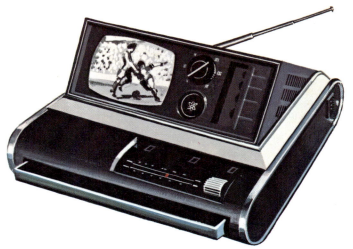

▲ **Portable Japanese television set** and radio combined, with a 5-in. (125-mm.) screen. It has a telescopic aerial and the screen folds away when

▼ **A vision of the future,** drawn by a French artist at the end of the 19th century. He foretold machinery combining the latest ideas in phonographs, cinema and telegraphy. These

not in use. The quality of the picture is excellent. Batteries and circuits are now made so small that a wrist-television may soon be available.

ideas seemed fantastic at the time, but in the United States there are now television sets which project flat-screen pictures onto a wall from a machine on the floor opposite.

How television developed

Still pictures had been transmitted over wires by Caselli, in France in the 1860s, but there was very little interest in the idea of transmitting *moving* pictures until, in 1884, a "scanning disc" was invented, which converted pictures in a camera into electrical signals. In 1897 Braun, in Germany, first made a cathode ray tube, but very few people saw its possibilities and 25 years later, at a meeting of the Radio Society of Great Britain, it was agreed that there was "not sufficient call for seeing by electricity to lead anyone to lay out the large sum of money which is necessary".

But they were wrong. Experimenters were at work in Germany and the United States, and a Scotsman, John Logie Baird, was beginning to attract attention. But Baird's mechanical method of scanning, by a perforated metal disc, did not produce clear pictures. Campbell-Swinton, also a Scotsman, thought out ways of using a beam of electrons as a scanner, but he died before his system was fully developed. In 1930 a successful camera tube was produced by a Russian, Zworykin, working in the United States, and he took out a patent for it.

The big electrical manufacturers were becoming interested in the idea of television and investing a lot of money in it. Baird's mechanical scanning method was subsequently dropped and since the late 1930s electronic systems have been used.

The first outside broadcast was in 1937; regular exchanges of programmes between European countries began in 1954; colour television was available in the 1960s; and video cassettes now enable viewers to reproduce recorded programmes. Space communications make it possible to watch almost perfect pictures of events all over the world as they are happening. Television is available in over 100 different countries and in some places, viewers watch the screens for an average of five hours a day.

the picture information allows the picture to be recreated by scanning the cathode-ray tube with electronic signals (4). In a colour receiver the picture and colour information are fed to a

special type of "shadow-mask" tube (5) with three electron guns which emit electrons. The shadow-mask is so arranged that electrons from the "red" gun always hit a phosphor dot

that produces red light. The same thing happens for green and blue. Many colour pictures are made up of more than a million of these tiny points of red, green or blue light. They

appear to our eyes as a continuous colour picture. Because electrons have virtually no weight they can be made to trace out a complete picture many times a second.

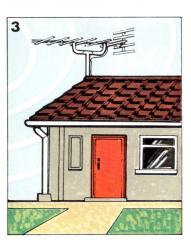

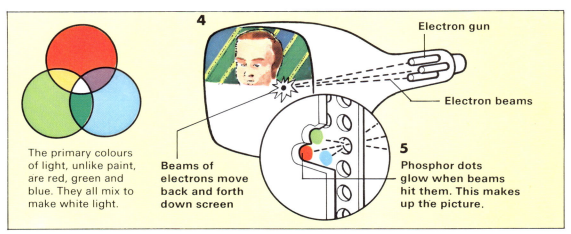

The primary colours of light, unlike paint, are red, green and blue. They all mix to make white light.

4 Beams of electrons move back and forth down screen

Electron gun

Electron beams

5 Phosphor dots glow when beams hit them. This makes up the picture.

"Firsts" of Home Inventions

An invention is not usually the work of one man and it is often impossible to say when a new idea was first thought of, or by whom. An idea sometimes grows in different places at the same time, without there being any direct connection between the people concerned.

Most ideas take time to grow before they are put into practice. Sometimes inventions are ready for use long before anybody is willing to invest money in making them; or an idea that seems of little use at first may later be usefully developed.

This list of inventions is selective – it consists of items which seem to the author to be interesting and to have had an important effect upon home life in the past 250 years. No list could be complete, but this one may help you to make your own.

1740	"Franklin" stove for heating (United States)
1778	Bramah's water closet (England)
1783	Leger invented flat-ribbon wick for oil burners (France)
1784	First scientifically-designed oil lamp by Argand (Switzerland)
1786	Earliest attempts at gas lighting in England and Germany
1790	Saint's sewing machine (England)
1795	Food canning by Appert (France)
1810	Krem's sewing machine (Germany)
	Sir Humphrey Davy produced carbon electric arc light (England)
1826	Walker's "friction lights" – matches (England)
1830	Thimmonier's sewing machine (France)
1836	Morse's electric telegraph (United States)
1845	Howe's sewing machine (United States)
1851	(Year of the Great Exhibition in London)
1855	Yale safety lock (United States)

1888	Berliner's gramophone demonstrated (United States)
1889	Bostel's "washdown" closet (England)
1890	Ash's refrigerator patented (England)
1891	Electric cooker demonstrated (England)
	First telephone link between England and Europe
	Zip fastener invented by Judson (United States), but first practical design in 1914
1892	Dyer's vacuum flask (England)
1893	Von Welsbach's improved incandescent gas mantle (Austria)
1894	Transmission of a message by "wireless telegraphy" by Sir Oliver Lodge (England)
1897	Braun's cathode ray tube for television (Germany)
1901	Marconi's wireless signals transmitted across the Atlantic
1904	Booth's domestic "suction machine" (England)

1855	First safety matches made (Sweden)
1856	Singer's domestic sewing machine (United States)
1860	Carré's refrigeration by ammonia (France)
	Still pictures transmitted by electricity by Caselli (France)
	Paraffin discovered in United States
1867	Shole's typewriter (United States)
1868	Maughan's gas geyser (England)
1874	First commercially successful typewriter (United States)
1875	Bell's "telephone": the first instrument to transmit sounds electrically (United States)
1877	Edison's first "talking machine" (United States)
1878	Swan's electric lamp (England)
	Edison's tinfoil phonograph demonstrated (United States)
1879	Edison's electric lamp
1888	Hertz (Germany) discovered electromagnetic waves, which led to Lodge's radio experiments in 1894

1904	Gillette's safety razor (United States)
	First gramophone disc to be recorded on both sides (Germany)
1911	Tungsten vacuum lamp made by Coolidge (United States), an improved version of the electric light bulb
	Carrier's work on air conditioning (United States)
1914	First washing machine with electric motor
	Sundback's zip patented (United States)
1921	First portable electric sewing machine
1923	Thermostat introduced for gas cookers
	First trans-Atlantic telephone link
1924	Aga cooker patented by Dr Dalen (Sweden)
	Domestic spin dryers on sale (United States)
	Baird's first TV camera for transmitting moving pictures (England)
1930	TV camera tube invented by Zworykin (United States)
1948	Slow-speed turntable, for long playing records (United States)
1959	Microwave oven patented (England)

Index

Note: Numbers in **bold** refer to illustrations

aerial, 42–3
Aga cooker, **15**
air conditioning, **4**, 12, **12**,
    see also heating
amplifier, 41
Argand burner, 6, **6**
Ash, 20
automatic exchange, 38
automatic tea-maker, **3**, 18, **18**

Baird, John Logie, 44–5
baths, 25–7, **25–7**
Bell, Alexander Graham, 38–40
Berliner, Emile, 41
boilers, 12–13, **12–13**, 26
Booth, H. Cecil, 29
Bostel, D. T., 24
"box"-iron, 18
box mangle, **30**
Bramah, Joseph, **22**
Branly, Professor, 42
Braun, Karl, 45
broadcasting, 42
Bunsen, Baron, 10, **19**
Burtt, William, 36

Campbell-Swinton, A.A., 45
candle, 4–6, **5–6**
carbon filament, 9
Carré, Ferdinand, 20
Carrier, Willis, 12
Caselli, 45
cathode-ray tube, 44–5, **44–5**
central heating,
    hot water, **13**
    Roman, 4, **4**, 10
    steam, 12, **13**
    solid fuel, **13**
    see also heating
cesspits, 22, **23**
chandelier, **6**
chimney, **4**, 5, **5**, 10, **10**
cholera, 22
cistern, **24**
Clayton, Rev., 7
clocks, 4, 35
clockwork,
    "jack", 14
    talking machine, **40**
Clos, Charles, 41
coal, 10, 12, 14
coal gas, see gas
coffee-maker, 18, **19**
colour television, 44–5, **44–5**
Cook, Colonel, 12
cooking,
    Aga, **15**
    Dutch oven, 14, **14**
    electricity, 16–17, **17**
    gas, 14, 16, **16**
    open fire, **5**, 14
    ranges, 14–15, **14–15**
    see also gadgets
Coolidge, 9

"crane" (chimney), **5**
crystal radio, 42
Cummings, Alexander, 22–3
cylinders (gramophone), 41

Dalen, Dr Gustav, 15
deep freezer, 20
de Forest, Lee, 42
dictating machine, 41
"digester", see pressure
    cooker
discs (gramophone), 40–1, **41**
disease, 22, 24
drains, see sanitation
Dutch oven, **14**

earth closet, 23–4, **23**
Edison, Thomas,
    electric light, 8–9
    phonograph, 40–1, **41**
    telephone, 38
egg beater, **19**
Egyptians, 4, 33
electricity, 5, see also
    heating, lighting, cooking,
    water heating
electro-magnetic waves, 42
electronic television signals,
    45
Ericcson telephone, 39
extractor fan, 18

factories, 5, 7, 22, 35
fan-heater, electric, **11**
fire, see heating, cooking,
    hearth
fireplace, see hearth
Fleming, Sir John, 42
flint, **5**
fluorescent, see lighting
food, see cooking
Franklin, Benjamin, 10, 12

gadgets, 18–19, **18–19**
gas, see lighting, heating,
    cooking, ironing, geysers
geysers, gas, **26–7**, 27
Gillette, King C., 32, **32**
girandole, **6**
Glidden, 36
gramophone, 40–1, **40–1**
graphophone, 40
Gray, Elisha, 38

Hammond's geyser, **37**
Harington, Sir John, 22–3
Harrison, James, 20
hearth, **4**, 5, **5**, 10, **10**
heating,
    district, 12
    electric, 10–12, **11**
    gas, 10, **10**, 12
    oil, 10, **11**, 12
    open fire, 4–5, **4–5**, 10, **10**
    Roman, 4, **4**, 10
    steam, 12, **13**
    see also air conditioning
Hertz, Heinrich, 42
hire-purchase, 34
His Master's Voice, 41
horn, 5, **5**

Howe, Elias, 35
hypocaust, 4, **4**

IBM "golfball" typehead, **37**
ice box, **20**
ice house, 20, **20**
ice store, **20**
immersion heater, 27
incandescent gas mantle, see
    lighting (gas)
Industrial Revolution, 5, 18
ironing, 18, **19**

"jack" spit, 14, **14**
Japan, 42, 45

kerosene, see oil
King, Alfred, 16
kitchen, see cooking, gadgets
"kitchener", 14, **15**
Krems, B., 35

labour-saving devices, 18–19,
    18–19
lamps, see lighting
lavatory, see water closet
lighting,
    candles, 4–6, **5–6**
    electric, 8–9, **9**
    fluorescent, **9**
    gas, 7–9, **7–9**
    neon, 8
    oil, 6–7, **6–7**
    rushlight, **5**, 6
    Roman, 4, **6**
    windows, 4–5
locks, 4, 33, **33**
Lodge, Sir Oliver, 42
loudspeakers, **41**

mangle, **30–1**
manual typewriter, 36
Marconi, Guglielmo, 42
Marconi-EMI television, **44**
Marconiphone radio, **43**
matches, 8–9, **8–9**
Maughan, Benjamin, 26–7
meat mincer, 18
mercury lamps, 8
Meucci, 38
micro-wave oven, 17
Middle Ages, 4–5, **4**
Mignon typewriter, **37**
Mill, Henry, 36
mincer, **18**
mixer, electric, **19**
mobile kitchen, 17
Morse code, 42
Morton's washing machine, **30**
Moule, Rev. H., 23
multi-point pressure geyser, **27**

neon, see lighting
nightman, **22**

oil, see lighting, heating,
    water heating

Pall Mall, London, 7
paraffin, see oil
patent, 14, 23

phonograph, 40, **40–1**
Pompeii, 4
Pratt's typewriter, **36**
Popov, Professor, 42
pre-historic man, 4
pressure cooker, **18**
Public Health Act, 1848, 22
pump, **5**

radiator, see central heating
radio, 5, 42–3, **42–3**
Radio Corporation of America,
    42
range, kitchen, 14–15, **14–15**,
    26
razor, safety, 32, **32**
record player, **41**
refrigerator, 20–1, **20–1**
Reis, Philip, 38
Remington typewriter, 36–7
Robinson, Thomas, 14
Roman homes, 4, **4**
Rothschild, 16
Rumford, Count, 10, **10**
rushlight, see lighting

safety razor, see razor
Saint, Thomas, 35
sanitation, 4–5, 22–4, **22–4**
sconce, **6**
Scott, Captain, 41
servants, 16, **17**, 18, 23, 25, 29
sewing machine, 34–5, **34–5**
Sholes, Christopher, 36–7
shower bath, **25**
Singer, Isaac, 34–5
solar energy, 27
spills, **5**
spin dryer, 30, **30**
spit, **5**, 14
standpipes, 25
steam, see heating
steam iron, **19**
stereo record player, **41**
"strip" lighting, see fluorescent
stove, 10–11, **11**, see also
    cooking
stylus, **41**
suction sweeper, see vacuum
    cleaner
Swan, Joseph, 8
syphonic water closet, 23, **23**

tape-recording, 40–1
taps, 25, **25**
tea-maker, **3**, 18, **18–19**
telegraph, 38, 42
telephone, 38–9, **38-9**
television, 5, 44–5, **44–5**
Telstar, 38
thermos flask, 33, **33**
thermostat, 17, 27
Thimmonier, 35
tinder box, **5**
toaster, **19**
transistors, **43**
tungsten filament, **9**
typewriters, 36–7, **36–7**

ultra-violet light, 9
utensils, 14, 18–19, **18–19**

vacuum cleaner, 28–9, **28–9**
vacuum pump, 9
video cassette, 45

Walker, John, 8
wash-basins, 25
wash-down closet, **23–4**
washing, *see* water supplies,
    baths
washing machine, 30–1, **30–1**
wash-out closet, **23**
water clock, 4
water closets, 22–4, **22–4**
water heating, 26–7, **26–7**
water supply, 4, **5**, 22–3, 25
wells, 5, 22
von Welsbach, 9, **10**
Wheeler and Wilson sewing
    machine, **35**
windows, 4–5
"wireless" telegraph, 42
wood, 10, 12

Yale lock, 33, **33**

zips, **32,** 33
Zworykin, Vladimir, 45